DESIGN AND PLASTICS

MIKE HALL

HODDER AND STOUGHTON
LONDON SYDNEY AUCKLAND TORONTO

CONTENTS

PREFACE

DESIGNING AND MAKING IN PLASTICS FOR CRAFT DESIGN TECHNOLOGY

The wide-ranging properties of plastics, their diverse and often exciting qualities, add sparkle to our everyday lives. Without plastics we would be unable to afford the many products we take for granted. Indeed many of the technological advances made by our society would not have been possible. Plastics are essential to the continuing development of our way of life through product design.

Craft Design Technology (CDT) seeks to develop technical and aesthetic awareness, originality, inquisitiveness and decision-making abilities through the experience of designing and making. Plastics play a major role in CDT. This book seeks to provide a foundation in plastics that will enable a young person fully to understand the qualities and potential of these exciting materials and enable them to make products that will fully exploit their properties. It covers, in considerable detail, the entire field of common plastics and attempts to demonstrate by practical examples how they can actually be used in designing and making objects. It aims to meet these needs over most of the CDT course from entry into secondary education at 11 through to 16-plus and GCSE requirements. Many topics barely touched upon in other school texts are treated in depth. The suggested projects and activities are designed to enable children to add to the text through personal investigation.

Plastics are the materials of our time and as such need to be understood and used to the benefit of all. Designing and making in plastics both fulfils our creative needs and educates through practical experiment and demonstration.

Acknowledgements

The author would like to express his appreciation to Jane Girdler and Jean Cawood for typing the manuscript and his son Richard for word processing the text in the early stages to establish the layout. A considerable debt is owed to David Shaw (ex CDT Advisor for Coventry) and to Campbell Grant and Martin Finn at Hodder and Stoughton for their technical and literary advice and to my colleagues, in particular Brian Bullock of the Department of Design and Technology at Loughborough University and Nick Way who read and corrected the manuscript and provided support and encouragement. His greatest debt however is to his students who over many years have taught him so much from their continued efforts to convert their product design ideas into plastics prototypes.

As a lecturer the author has gained much technical advice from sources in the Plastics Industry and many reference works providing the detail necessary for this book and is deeply indebted to the originators of this information. In particular the tables shown on page 123 which are reproduced with the permission of SCDC Publications from Design with Plastics (Project Technology Handbook), (Heinemann Educational Books 1974) and Plastics Engineering by R. J. Crawford (Pergamon 1981/83). The cover photograph is reproduced by kind permission of the Design Council.

Michael J. D. Hall 1988

ISBN 0 340 40528 7

First published 1988

Copyright © 1988 Mike Hall

Typeset by Graphicraft Typesetters Ltd, Hong Kong
Printed in Great Britain for Hodder and Stoughton Educational a division of Hodder and Stoughton Ltd, Mill Road, Dunton Green, Sevenoaks, Kent by Butler & Tanner Ltd, Frome and London

INTRODUCTION: THE IMPORTANCE OF PLASTICS

The large range of plastics materials in use today have been discovered and developed by scientists during the last 120 years. They have been discovered because man is constantly searching for new ways of making things and making them longer-lasting, cheaper, lighter and better able to meet the needs of new technology.

Plastics are not cheap materials, but because they can be processed more efficiently on automated machines they are cheaper than many others after manufacture.

Today most people want a car, stereo, radio, TV set, video recorder, and all the labour-saving devices to be found in our homes. Plastics play an important part in all of them. Without these materials it would not be possible to make many of these products, and certainly not at a price we could all afford.

Outside the home, too, plastics are crucial to innumerable activities. If all plastics products were to vanish overnight, the effect on us and the way we live would be catastrophic. Virtually everything everywhere would stop. Plastics are essential to our computers, communications systems, leisure equipment, food industry, and medical and other services. Almost everything you can think of these days has at least one part made from plastics materials.

Devices incorporating plastics perform many different functions and plastics provide a wide variety of suitable properties to choose from. It is therefore important to choose the correct material for the job. Some plastics are much more expensive than others.

What are plastics? They are man-made substances that may be shaped under heat and pressure but are stable in normal use. There are two main types of plastics commonly used for making the products in our homes. They are called *thermoplastic* and *thermosetting plastics*. They are different in that: (a) the methods used to process them are different; and (b) thermosetting plastics ('thermosets') change chemically during processing, whereas thermoplastics do not. Tables 1 and 2 show common plastics, their applications and generic names. The common names are used to identify materials throughout the book.

Table 1 Thermoplastics

Common name	Old generic name	New generic name	Applications
polyethylene (polythene, PE)	polyethylene	poly(ethene)	washing-up bowls
high-density (HDPE)	high-density (HDPE)	high-density	milk-bottle crates
low-density (LDPE)	low-density (LDPE)	low-density	carrier bags
PVC (vinyl)	polyvinyl chloride	poly(chloroethene)	floor covering, rainwater guttering, shower curtains, upholstery fabrics
polystyrene (styrene, PS)	polystyrene	poly(phenylethene)	tape cassette boxes
polypropylene (propylene, PP)	polypropylene	poly(propene)	injection-moulded chairs
ABS	acrylonitrile butadiene styrene	poly(propenenitrile) buta-1, 3-diene poly(phenylethene)	suitcases
Acrylic (Lucite, Perspex, Oroglas, PMMA)	polymethylmethacrylate	poly(methyl 2-methyl propenoate)	car rear lights
SAN	styrene acrylonitrile	Poly(phenylethene propenenitrile)	translucent coloured drinking glasses
CAB (cellulose acetate butyrate)	CAB (cellulose acetate butyrate)	poly(cellulose ethanoate butanoate)	street signs
cellulose acetate (acetate, CA)	cellulose acetate	poly(cellulose ethanoate)	photographic film, spectacle frames
cellulose nitrate (Celluloid)	cellulose nitrate	poly(cellulose nitrate)	table tennis balls
nylon	polyamide		gears, curtain fittings
polycarbonate (Lexan, PC)	polycarbonate	poly(carbonate)	camera bodies, car bumpers
PET	polyethyleneterephthalate	poly(ethene polyester benzene-1, 4-dicarboxylate)	pressurised soft-drinks bottles, draughtsman's drawing film
acetal	polyacetal	poly(1, 1-diethoxyethane)	taps
EVA	ethylene vinyl acetate	poly(ethene ethenylethanoate)	lifejacket buoyancy foam
PTFE	polytetrafluoroethylene	poly(tetrafluoroethene)	non-stick pan coatings

THERMOPLASTICS

During processing the chemical composition of thermoplastics remains unchanged, but as the plastics are heated, the minute particles, or

Products made before plastic came into general use.

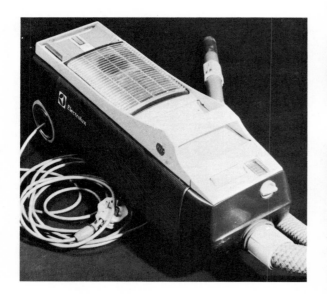

The parts of this domestic appliance have been injection moulded in ABS (acrylonitrile butadiene styrene).

Cellulose nitrate table tennis balls.

Castor and gears in nylon.

Car rear light injection moulded in acrylic.

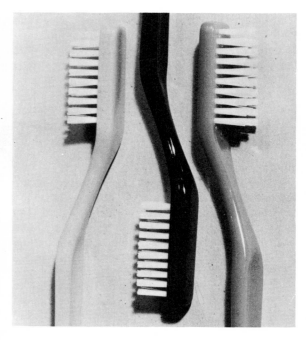

Cellulose acetate toothbrushes.

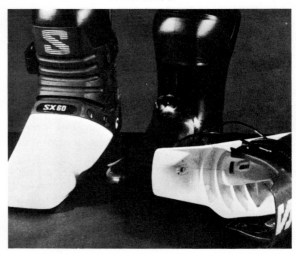

Injection moulded ABS ski boots.

4

Injection blow moulded lemonade bottle in PET
(polyethyleneterephthalate).

Extrusion blow moulded bottle in high density
polyethylene.

Doll rotationally cast in PVC plastisol.

Motor cyclist's crash helmet injection moulded in
polycarbonate.

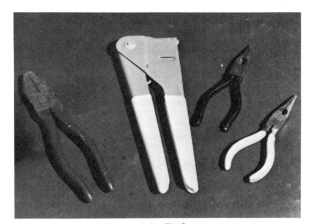

Tool handles dip-coated in PVC.

◀Jug and beakers injection moulded in SAN (styrene
acrylonitrile).

Urea formaldehyde compression mouldings.

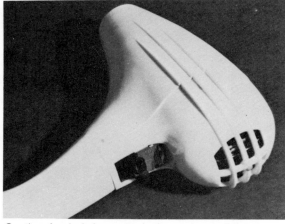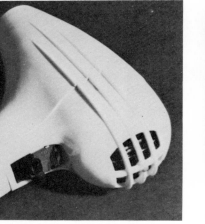

Casting for hairdryer made by compression moulding in urea formaldehyde.

Sea fishing reel moulded in glass reinforced polyester.

Melamine formaldehyde tableware made by compression moulding.

Phenol formaldehyde compression moulded bowl.

Sandwich toaster with phenol formaldehyde (bakelite) compression moulded, heat resistant handles and a non-stick coating of ptfe (polytetrafluoroethylene).

Clock parts embedded in polyester casting resin.

molecules, (see page 8) vibrate and become free to move past each other, so making it possible to shape the material. In some processes the plastics are heated until they become syrupy fluids. Then, when great pressure is applied, they can be forced through passageways into spaces (mould cavities) where they are cooled and become solid, taking on the exact shape of the space. Each mould cavity makes a new product or component.

Thermoplastics can be bought in the form of powders or granules, or as sheets, rods, tubes or sometimes as liquids. They can be heated and moulded by a variety of methods. Some materials are more suited to one process than another but in all cases the principle is the same: forming takes place with heat and pressure but chemically the material remains the same.

THERMOSETTING PLASTICS

A thermosetting plastics material is one that changes chemically during the moulding process. Usually it is formed by heat and pressure, but once this chemical change has taken place the new material cannot be reshaped by further heat and pressure. Some materials may be cast at room temperature and atmospheric pressure and hardened by the addition of chemical agents.

In school work very few thermosetting compounds are used. Polyester resin is the most important example: it is the most common material for bonding with glass fibre to make glass fibre-reinforced products such as canoes, chairs and many other items that consist essentially of thin shells with one good surface and high strength.

Plastics offer a wide range of materials for design. All work in this field demands planning and forethought, careful materials selection and meticulous processing of both moulds and finished products. A high standard of workmanship is essential, and when it is coupled with sensitive design, it will result in a high-quality product.

You must always base the selection of materials and processes on the uses to which the finished product will be put and the facilities and time available. Where possible this book indicates the applications of each process described and the materials required. Detailed methods of mould-making to suit the various processes are given so that you can see how to approach the design and plan a project. You should be able to think both about the product in its finished form and about the mould, which is its opposite or 'negative'.

Plastics materials can be used to solve many construction problems but they cannot be expected to do everything. There are many occasions when you have to use metals and wood for structural purposes, and you are encouraged to use the best material for the job. This is an essential part of good design.

Table 2 Thermosetting plastics

Common name	Old generic name	New generic name	Applications
phenolic (PF, Bakelite)	phenol formaldehyde	poly(phenol methanal)	pan handles, insulation foams
urea (UF)	urea formaldehyde	poly(carbamide methanal)	light switches
melamine (MF)	melamine formaldehyde	poly(melamine methanal)	cups, kitchen workbench laminates
polyester	polyester		GRP car bodies, canoes
epoxy	epoxy		commercial tooling, electrical casting medium, adhesives, coatings
polyurethane (PU, urethane) (Note: some of these are thermoplastics, depending on the mix.)	polyurethane flexible		upholstery foam, skateboard wheels, some coated fabrics, shoe soles
	polyurethane rigid		insulation foam, structural furniture foam, adhesives, coatings
synthetic rubbers neoprene rubber	polychloroprene	poly(2-chlorobuta-1, 3-diene)	divers' wet-suits, inflatable dinghies
butyl rubber			inner tubes for tyres
silicone rubber			medical, food-processing (for linings, seals, etc.)

THE RAW MATERIALS OF PLASTICS

Coal, oil, plants and milk are the most common raw materials from which plastics are made.

Coal and oil provide the main carbon compounds that form the chemical backbone for almost all plastics. Coal and oil are used for many purposes, and less than five per cent is consumed in the manufacture of plastics.

Plants provide cellulose, which is made into a variety of sheet thermoplastics, such as cellulose nitrate (for table-tennis balls etc.), cellulose acetate (for spectacle-frames etc.) and cellulose acetate butyrate (for some street signs etc.). Sugar can provide pure carbon and, being fast-growing, can be used as a renewal resource for plastics.

Milk provides casein, which, when chemically treated and compressed, forms another group of plastics.

CHEMICAL STRUCTURE OF ORGANIC COMPOUNDS

Most compounds based on the element carbon are said to be *organic*. An element is a substance that cannot be broken down into any simpler substances, and therefore its atoms are all of the same kind. An *atom* is the smallest part of an element that can take part in a chemical reaction. Hydrogen, oxygen and chlorine are all examples of elements.

A *compound* is made up of atoms of different elements, chemically combined in a particular ratio; for example, water has two hydrogen atoms combined with one oxygen atom. Different compounds result if the same elements are combined in different proportions; for example, hydrogen peroxide has two hydrogen atoms and *two* oxygen atoms combined. Identical atoms

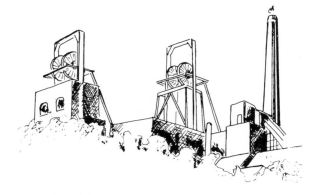

can also join together; oxygen, for example, normally consists of pairs of oxygen atoms linked together.

These combinations of atoms are *molecules*, which are the smallest units of elements or compounds that can exist by themselves.

Chemists are able to make atoms combine with each other in special relationships to produce plastics. They have many ways of controlling the number and arrangement of the different atoms in the molecule of a compound.

An atom of hydrogen can form one link with an atom of carbon or another substance. An atom of oxygen can form two links, an atom of carbon four links. A link between atoms is known chemically as a *bond*. Carbon can make a greater number of chemical substances than can all the other elements combined.

Carbon atoms can link together to form chains or rings. And one kind of ring, known as a *benzene ring* can be attached at the side of a carbon chain. (See page 13.)

Sometimes one carbon atom forms two links with another carbon atom. The compound is called *unsaturated*, meaning that one of the two links can be broken in order to make a different molecule. When all the links between atoms are single, a *saturated* chemical compound is

created. In this case the molecule cannot be altered by normal chemical means. Usually a double bond can be broken easily, and when this happens long chain molecules can be formed.

Polymerization

A basic unsaturated organic molecule containing double carbon-to-carbon bonds, is called a *monomer* (mono = one). When the double bonds are broken to form a long chain molecule, the new material is said to be a *polymer* (poly = many). This process is known as addition polymerization, because it consists of the addition of molecules of the same monomer to form chains of perhaps 10,000 carbon atoms. (See page 9.)

Another polymerization process is possible, involving the use of two or more monomers of different substances. The product of this reaction may be a long chain molecule, or a complex, cross-linked three-dimensional structure; usually water or hydrogen chloride is formed as a by-product, giving rise to the name condensation polymerization for the reaction. In this case the water or hydrogen chloride is the condensate.

Addition polymerization creates long chain molecules that have thermoplastic properties. These molecular chains lie on top of and alongside each other in a jumble, lightly held together by weak attractive forces known as van der Waals forces. Heat weakens these forces, very noticeably in the normally rigid thermoplastics material, which becomes soft and floppy. The heat permits the molecules to slide past one another, making the material mouldable under pressure. When the material has cooled, the van der Waals forces regain their strength and hold the molecules in their new positions. This heating and cooling can take place more than once, so that a thermoplastic can be shaped and re-shaped many times.

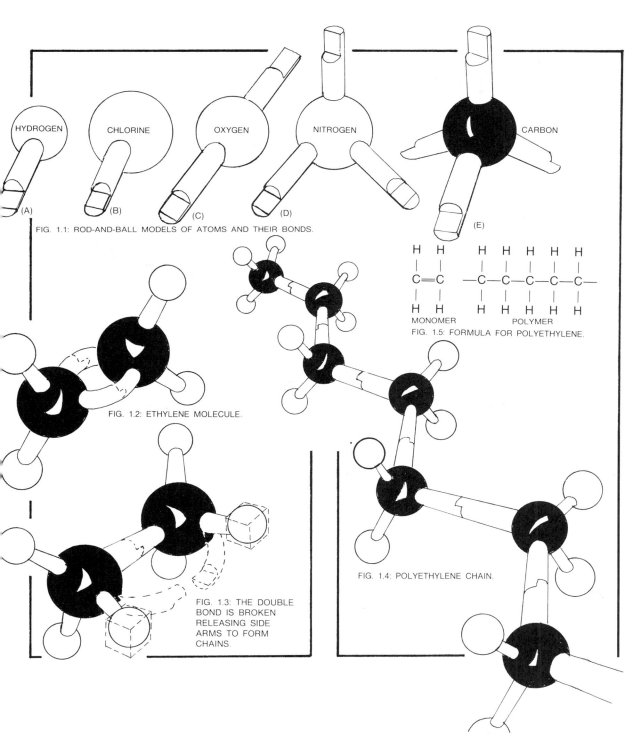

FIG. 1.1: ROD-AND-BALL MODELS OF ATOMS AND THEIR BONDS.

HYDROGEN   CHLORINE   OXYGEN   NITROGEN   CARBON

(A)   (B)   (C)   (D)   (E)

FIG. 1.2: ETHYLENE MOLECULE.

FIG. 1.3: THE DOUBLE BOND IS BROKEN RELEASING SIDE ARMS TO FORM CHAINS.

FIG. 1.4: POLYETHYLENE CHAIN.

$$\underset{\text{MONOMER}}{\overset{\begin{array}{cc} H & H \\ | & | \\ C = C \\ | & | \\ H & H \end{array}}{}} \quad \underset{\text{POLYMER}}{\overset{\begin{array}{ccccc} H & H & H & H & H \\ | & | & | & | & | \\ -C-C-C-C-C- \\ | & | & | & | & | \\ H & H & H & H & H \end{array}}{}}$$

FIG. 1.5: FORMULA FOR POLYETHYLENE.

Thermosetting plastics have a cross-linked structure that, when cured either chemically or by heat, forms a group of fixed atoms held firmly in relation to one another. These atoms will not separate when reheated — they can be heat- and pressure-moulded once, or chemically cured once, but after this they cannot be re-moulded.

Thermosetting plastics mentioned in this book are rigid, but different properties can be achieved by modifying the chemicals used in the curing process. Once cured, reshaping can only be achieved by cutting or machining.

THERMOPLASTICS

Polyethylene

Figures 1.1 A–E on this page show the main atoms found in many plastics materials and the number of 'arms' that each atom has. Hydrogen and chlorine, with one each, can attach themselves only to one other atom, while oxygen, nitrogen and carbon can link with two or more other atoms. Each arm is shown with a half-joint to suggest that it must 'mate up' with a similar half-joint on a spare arm of another atom. Figure 1.2 shows a molecule of ethylene (an unsaturated compound) with two carbon atoms linked by a *double bond* (two arms joined). When this double bond is broken (Figure 1.3), many of these ethylene molecules can join up to form one very long chain. Figure 1.4 shows the end of such a chain, and since there are many 'ethylene' molecules in it, it is now called 'polyethylene'.

When polyethylene is made by the industrial chemist, large numbers of the chains are created, which lie entangled like spaghetti in tomato sauce. When cold these chains stick to one another like a solid block of spaghetti; when warm they are free to slide around, until they are cooled again, when they 'set' in another shape. The van de Waals forces of attraction act like the tomato sauce in holding the long chains together.

Figure 1.5 shows a diagrammatic representation of the ethylene monomer, and the polymer formed after the double bonds have been broken and linked up to form a chain.

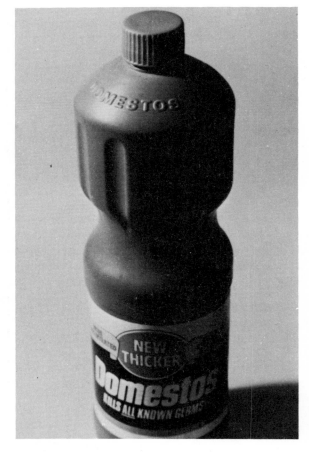

High density polyethylene material, extrusion blow moulded to form a strong container for domestic chemicals.

Low density polyethylene has been screen printed and welded to make these carrier bags.

High density polyethylene has been injection moulded to make this washing-up bowl.

Injection moulded chair made from polypropylene.

Polypropylene 'integral' hinge.

Polypropylene

Polypropylene, or polypropene, (Figure 1.6), is a very similar material to polyethylene, in appearance, touch and many other properties. It is a member of the same family and has almost the same chain structure, except that on every other carbon atom one of the hydrogen atoms is replaced by another carbon and three hydrogen atoms. These are always arranged in the same pattern and, because they stick out, tend to interfere with neighbouring chains.

Figure 1.7 shows the propylene monomer in diagrammatic form, with the single unsaturated double bond, which, when broken, allows the molecules to link up to form the long chain molecule of polypropylene.

Polypropylene string.

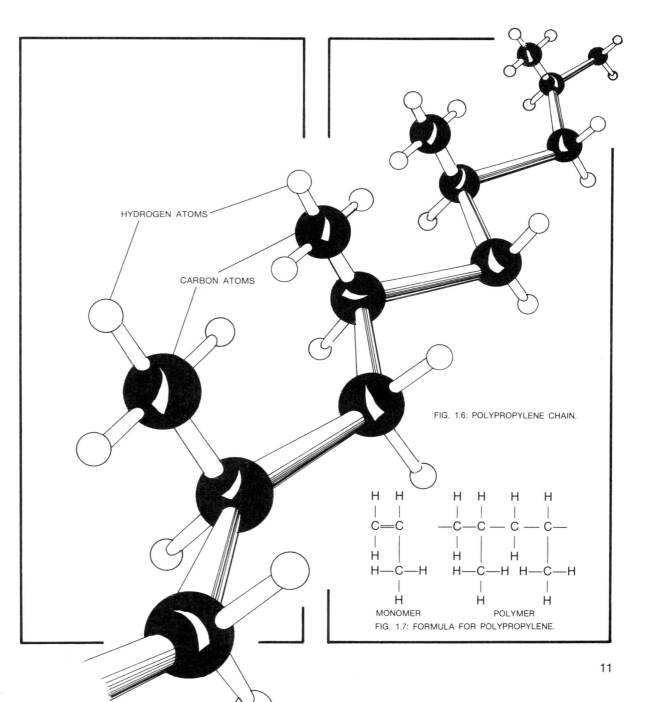

HYDROGEN ATOMS

CARBON ATOMS

FIG. 1.6: POLYPROPYLENE CHAIN.

MONOMER   POLYMER

FIG. 1.7: FORMULA FOR POLYPROPYLENE.

11

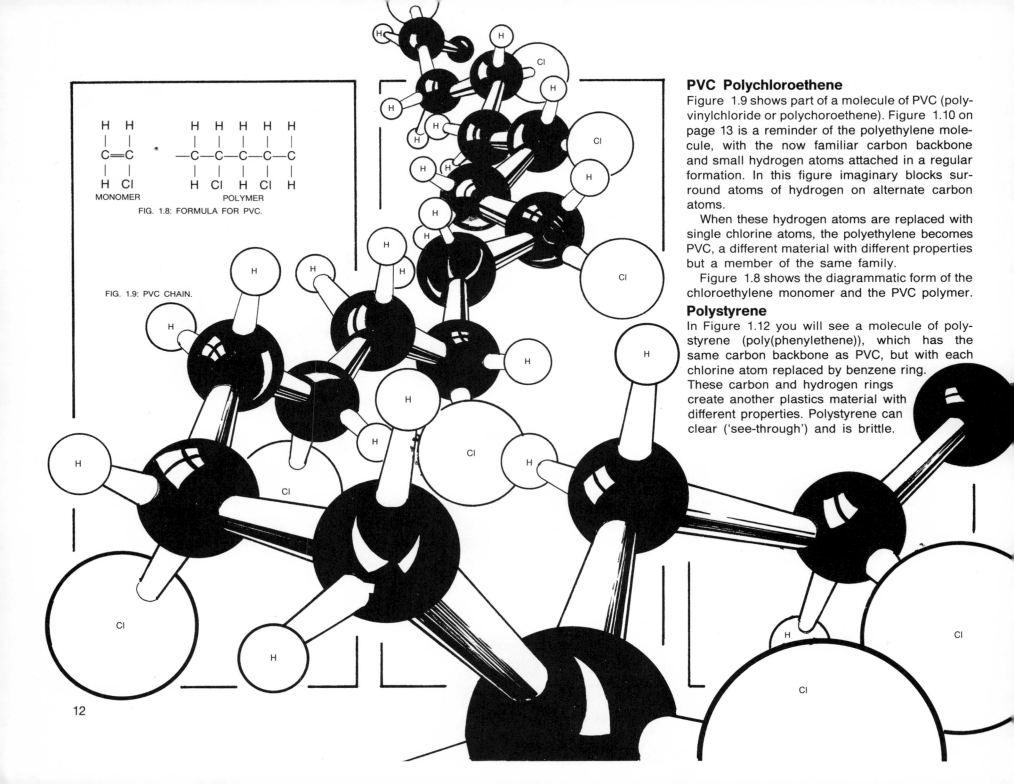

PVC Polychloroethene

Figure 1.9 shows part of a molecule of PVC (poly-vinylchloride or polychoroethene). Figure 1.10 on page 13 is a reminder of the polyethylene molecule, with the now familiar carbon backbone and small hydrogen atoms attached in a regular formation. In this figure imaginary blocks surround atoms of hydrogen on alternate carbon atoms.

When these hydrogen atoms are replaced with single chlorine atoms, the polyethylene becomes PVC, a different material with different properties but a member of the same family.

Figure 1.8 shows the diagrammatic form of the chloroethylene monomer and the PVC polymer.

Polystyrene

In Figure 1.12 you will see a molecule of poly-styrene (poly(phenylethene)), which has the same carbon backbone as PVC, but with each chlorine atom replaced by benzene ring. These carbon and hydrogen rings create another plastics material with different properties. Polystyrene can clear ('see-through') and is brittle.

MONOMER   POLYMER

FIG. 1.8: FORMULA FOR PVC.

FIG. 1.9: PVC CHAIN.

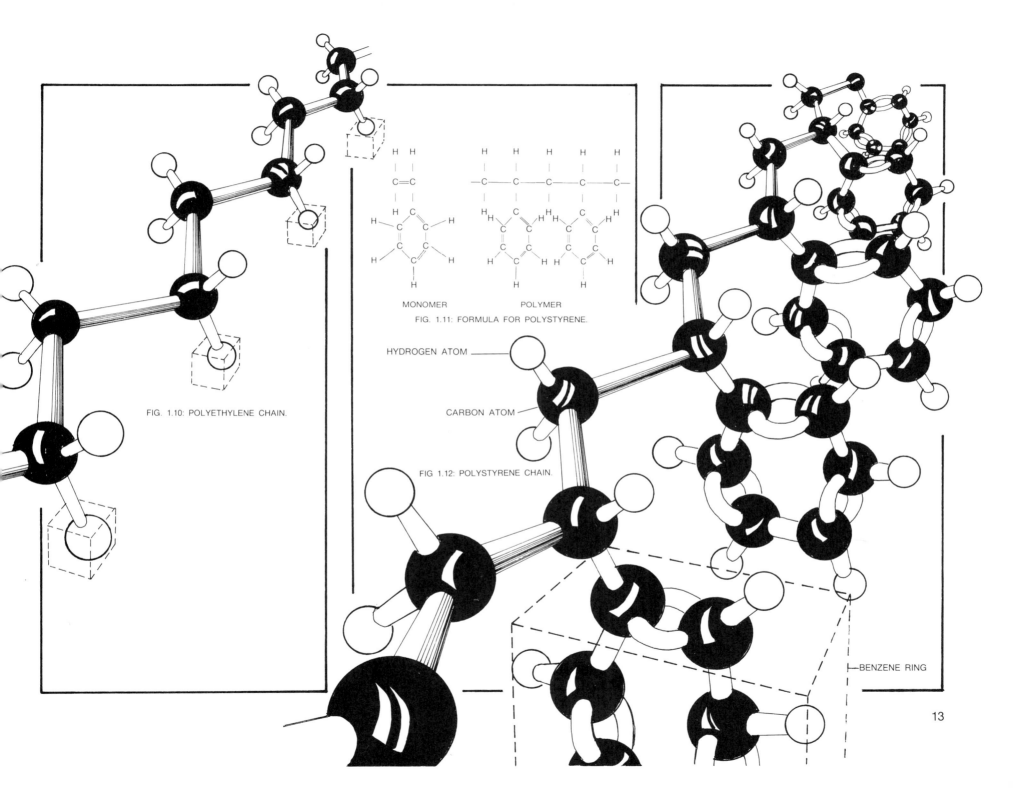

FIG. 1.10: POLYETHYLENE CHAIN.

H   H   H   H   H   H   H
| | | | | | |
C = C   — C — C — C — C — C —
| | | | | | |
H   C   H   C   H   C   H

MONOMER   POLYMER

FIG. 1.11: FORMULA FOR POLYSTYRENE.

HYDROGEN ATOM

CARBON ATOM

FIG 1.12: POLYSTYRENE CHAIN.

BENZENE RING

13

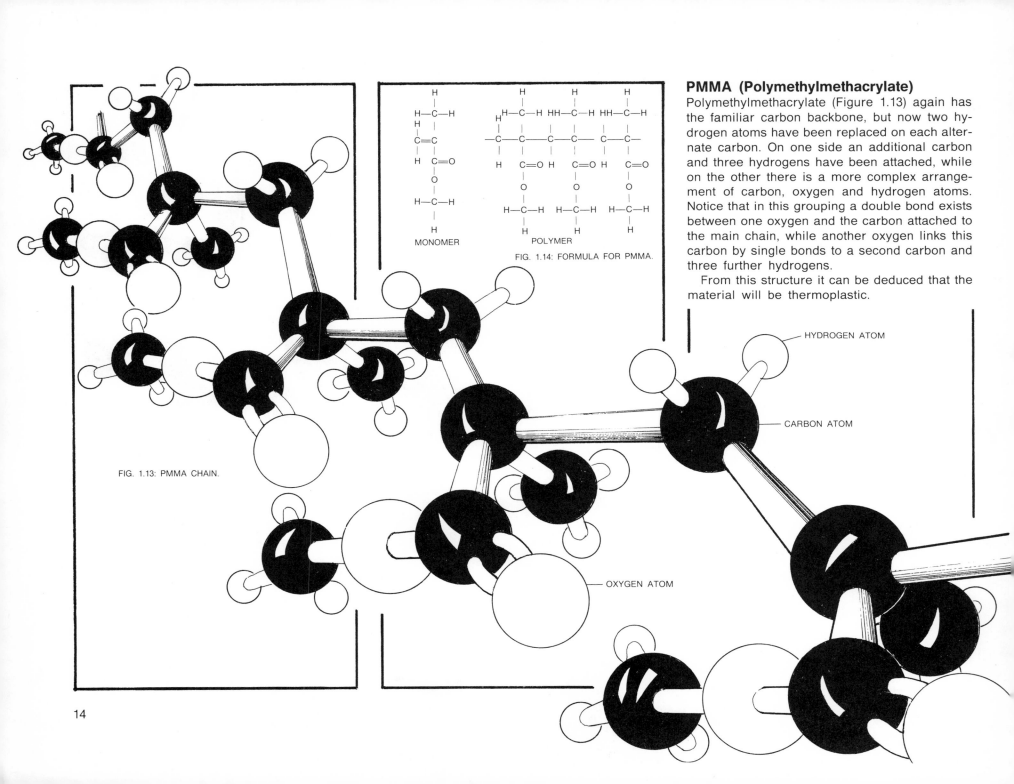

PMMA (Polymethylmethacrylate)

Polymethylmethacrylate (Figure 1.13) again has the familiar carbon backbone, but now two hydrogen atoms have been replaced on each alternate carbon. On one side an additional carbon and three hydrogens have been attached, while on the other there is a more complex arrangement of carbon, oxygen and hydrogen atoms. Notice that in this grouping a double bond exists between one oxygen and the carbon attached to the main chain, while another oxygen links this carbon by single bonds to a second carbon and three further hydrogens.

From this structure it can be deduced that the material will be thermoplastic.

MONOMER

POLYMER

FIG. 1.14: FORMULA FOR PMMA.

FIG. 1.13: PMMA CHAIN.

HYDROGEN ATOM

CARBON ATOM

OXYGEN ATOM

THERMOSETTING PLASTICS

Urea formaldehyde

Figure 1.15 shows a large molecule formed from five molecules of urea and four molecules of formaldehyde. This is the product of a condensation reaction between two dissimilar types of molecule.

It is important to notice that the shape of the urea formaldehyde molecule is unlike any of the others previously shown. It links three-dimensionally, and the schematic drawing shows it spread out in a box to suggest its arrangement in space.

Imagine many more combined together and tightly packed. They would become interwoven and inseparable, like twigs and branches in an old hedge, forming one giant molecule. So it is with all thermosetting plastics. The atoms become chemically linked and cannot be re-shaped because they will not separate and slide on heating. In effect, when they are processed to make a product the heat causes a chemical reaction to take place, and the pressure compacts the atoms into a form that is then permanent. The new material cannot easily be attacked by other chemicals, is unaffected by heat and when properly moulded makes strong, long-lasting, hard-surfaced components.

Figure 1.16 shows diagrammatically the arrangement of the molecules of urea and formaldehyde before and after combining to form the new material. Examine Figure 1.16 and find where the molecules start and finish on Figure 1.15. Notice how the molecule in the centre is an incomplete bridge between the other four. The formaldehyde carbon and two hydrogen atoms form links between the molecules and are shared.

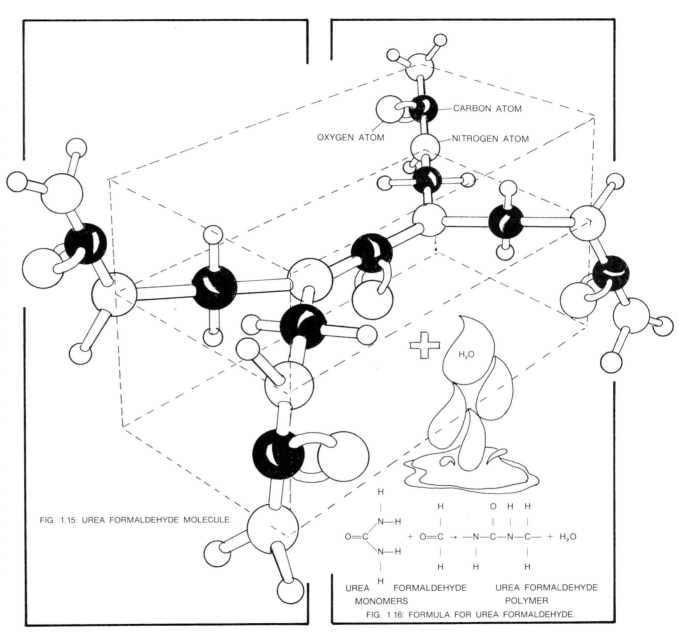

CARBON ATOM

OXYGEN ATOM   NITROGEN ATOM

H_2O

FIG. 1.15: UREA FORMALDEHYDE MOLECULE.

UREA   FORMALDEHYDE   UREA FORMALDEHYDE
MONOMERS   POLYMER

FIG. 1.16: FORMULA FOR UREA FORMALDEHYDE.

PROJECTS AND ACTIVITIES

(1) Find a steel ruler, a wooden ruler and a plastic ruler of similar size. They are samples of three different materials. Devise some tests to show how the materials differ. Write down a list of the properties of each material. Then using this information decide which materials are most suitable for making:
(a) a writing instrument;
(b) a bath toy;
(c) a door handle;
(d) a pair of scissors;
(e) a bottle to contain hair shampoo;
(f) a garden chair.

(2) Some materials that occur naturally show thermoplastics properties when heated (hair, bone, horn, tar and pitch for example). Can you describe how we use these properties and find any examples of uses for these materials?

(3) Look around your home and find an example of a product with variations made from different materials (a coathanger, for example). Collect three versions made in different ways from different materials. Draw them in detail and discuss their structural, practical and visual qualities.

(4) Why are van de Waals forces important in thermoplastics materials and what effect does heat have on them?

(5) In the kitchen you will find storage containers made entirely from plastics materials. Cooking pots and pans are made from glass, metal and ceramics but not plastics. Why is this?

(6) Can you find out how many parts of the human body can be replaced with plastics materials? Why are plastics suitable for replacement parts? Draw a chart of a human figure, identify where the replacement parts are used and label them with the names of the plastics.

(7) Plastics are used as substitute materials for some natural materials. Can you find and make a display of some examples of plastic that look and perhaps feel like (a) leather, (b) bone, (c) ivory, (d) marble, (e) string, (f) silk, (g) fur, (h) wool, (i) wood, (j) chrome metal. Discuss in class whether it is good practice to make plastics look like natural materials.

(8) Some people claim that plastics packaging has increased the litter problem in this country. Is it because plastics do not rot easily that the problem arises or is it for another reason? See if you can find out anything about **bio-degradable** plastics (plastics that rot and breakdown by the action of bacteria and weather). Would these materials solve the litter problem or would other problems occur?

(9) Take several pieces of different 'rigid' plastics materials, each 110 mm long, 20 mm wide and 3 mm thick. Clamp them firmly to the end of a bench, leaving 100 mm of each sticking out. Suspend a 10 g weight from each and measure the amount of deflection (sag) that takes place. Leave them in position with their weights for two weeks and measure the deflection every day. Draw a graph for each material. This will show you the property of creep. At the end of the period examine and record any permanent 'set' that the materials have taken. How important is this in designing plastics products?

(10) Can you find the following names of plastics materials in the word search below? (Letters can be used more than once.)

ACRYLIC
ACRYLONITRILE BUTADIENE STYRENE
CELLULOSE ACETATE BUTYRATE
EPOXY
MELAMINE FORMALDEHYDE
POLYAMIDE
POLYCARBONATE
POLYESTER
POLYETHENE
POLYPROPYLENE
POLYSTYRENE
POLYURETHANE
POLYVINYLCHLORIDE
PHENOL FORMALDEHYDE
UREA FORMALDEHYDE

E	F	O	R	M	A	L	D	E	H	Y	D	E	O	P
N	J	E	L	I	R	T	I	N	O	L	Y	R	C	A
I	E	B	P	O	L	Y	U	R	E	T	H	A	N	E
M	D	U	R	E	S	O	L	U	L	L	E	C	Y	N
A	I	T	E	A	C	E	T	A	T	E	B	A	L	E
L	M	A	T	Y	E	P	O	X	Y	K	U	R	O	L
E	A	D	S	P	O	L	E	L	Y	V	T	B	P	Y
M	Y	I	E	L	L	E	N	T	L	I	Y	O	H	P
A	L	E	Y	X	P	O	L	Y	O	N	R	N	E	O
C	O	N	E	N	E	H	T	E	P	Y	A	A	N	R
R	P	E	S	T	Y	R	E	N	E	L	T	T	O	P
Y	U	S	E	D	I	R	O	L	H	C	E	E	L	Y
L	R	E	D	Y	H	E	D	L	A	M	R	O	F	L
I	E	E	E	N	E	R	Y	T	S	Y	L	O	P	O
C	A	F	O	R	M	A	L	D	E	H	Y	D	E	P

2 PLASTICS DESIGN

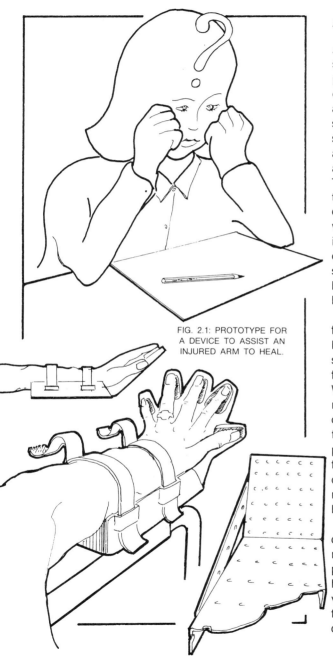

FIG. 2.1: PROTOTYPE FOR A DEVICE TO ASSIST AN INJURED ARM TO HEAL.

DESIGN PROJECTS

A teacher may suddenly ask you to think up something to design and make as a project. It is probable that your first thoughts will be. 'What do I want .....?' and you will start to consider adapting an existing product. You will then make sketches of how it could look, choose one design, and go on to make an engineering drawing and perhaps detailed component drawings. Finally you will make the article you have designed. This is fine: you will solve many problems along the way in the choice of materials and processes, of shapes, sizes and colours, and you will grapple with costings. Your design will be individual and probably different from the product that first gave you your idea. However, you started with a pre-existing solution to your problem, and that solution has prevented you from being really original.

Here is another situation. An old friend of the family has had a heart attack, and you go to visit him in hospital. You find that he has lost strength and control in his left arm, and his fingers have closed into the palm of his hand. While talking to him you learn that the hospital needs a device that will keep his arm outstretched and his hand open. If a device was made that could do this he would be able to have physiotherapy to bring back some strength to the muscles in his arm and hand. However, a device does not exist, so it is probable that the muscles will waste away and his arm will become useless.

To solve this problem, it is necessary first to check that this is the true situation. If it is, you need to discuss the type of device that the physiotherapist believes is necessary. She will have partly formed ideas based on other devices used in the hospital. She may also tell you that it must fit on one or more types of wheelchair, be used for so many hours per day for

several months, be adjustable and easy to clean, look like a comfortable arm rest, and so on. She may do a rough sketch of what she has in mind, but it will not be a full design.

You measure the wheelchairs that the device has to fit, and your friend's arm and hand. When you get home you start to sketch your ideas but find you can solve only a few of the problems. You get out your construction set — a Meccano set, perhaps — and make a full-size model. Now you can see 'all round' the problem.

You find a piece of foam to pad over the model and then look at ways for stretching out the fingers. By laying your own arm on your model you realize that it will fall off the device if not held by some sort of strap (Figure 2.1). The strap must be adjustable and capable of being undone by a person with one hand.

The product slowly designs itself as you discover the real problems and solve them. Your model will probably look very crude at this point. However, your teacher can tell you about the materials and processes you can use to change your model into a more realistic prototype. You prepare some drawings and make the prototype, using some of the methods discussed in this book. Then you take it to the hospital to test it. Your friend wants changes, the physiotherapist wants changes, and you've already thought of several changes of your own. You go back to the drawing-board, redesign the device and build prototype 2, a greatly improved version.

By the time you have finished you have provided someone with hope for a better future, you have solved a real problem and you have learned a great deal. You will have thought, researched, experimented, tested and applied your knowledge and skill to a pressing practical problem.

These two types of design situation are both common and both lead to the development of new products. The second satisfies a real need and usually demands a broader approach. The first type of project can often be merely change for the sake of change. For someone with only limited experience this usually leads to an increase in knowledge and skill in the handling of materials and processes, and in control over visual form. Quite often such a designer works within the limits of his own ideas without contact with other people. When this happens he is likely to juggle things around to suit himself rather than work within the constraints imposed by specialists, requirements, or by circumstances. It often requires a person with a lot of experience to get the best out of such a situation.

DEVELOPING A NEW PRODUCT

Below we list a fairly standard sequence of steps that you will probably have to go through in the course of designing and making a new product. You can use it as a checklist to make sure you don't overlook some vital step when planning the work, or when it has begun. But you may well have to depart from it in particular cases, and not all the steps will always be necessary.

The brief
Write it out clearly.
State the problem, rather than specifying the product you're aiming at, which will limit your imagination.
Specify what costs are acceptable.

Research
Study the activity in which the problem arises, the people who will be using your product, and the environment in which it will be used.
Study possible materials and methods of moulding.
Look at existing products that are used to aid this activity.
Get relevant ergonomic information — that is,

information relevant to the efficient carrying out of the activity, such as measurements of reach, or thrust, or frequency of use, or users' bodily dimensions.
Find out if there is a relevant British Standard.

Specification
Prepare a technical specification outlining the performance required.
Decide, as far as possible at this stage, what materials and processes could be used. (See charts on pages 124 and 125).
Make tests on materials and processes if necessary.
Estimate costs and confirm that they are within your limits.

Design
Sketch possible shapes for the product.
Sketch details, joints, etc, and methods of fixing typical sections, and possible assembly methods.
Decide on the most suitable design.
Prepare a general arrangement drawing.
Make a cardboard model — full-size if possible, otherwise to scale.
Test the model if possible.
Modify it if necessary.
Make detailed drawings of mouldings and moulds.
Show how mouldings are to be assembled.

Making
Make moulds.
Make mouldings, trim to shape and fit components.
Assemble complete product.

Test and evaluate
Test the product, using a typical operator in the proposed working environment.
Photograph the product in use.
Prepare a critical report.

Redesign
Suggest areas where the product needs to be changed, make drawings to show improvements, and repeat as much of the sequence as necessary from the design stage.

IDEAS FOR PROJECTS

Below is a list of project areas, starting with portable containers. Looking through it, you may recognize something that could solve a problem that you have.

For example: suppose your dad has a beach-casting sea-fishing rod that will not fit into your car. When you go on holiday the rod has to be strapped to the car roof-rack. When not in use the rod is constantly a worry because your dad is afraid it will be stolen. Suppose you bought a suitable length and diameter of plastic rainwater pipe and designed fittings and fixings so that it could be locked and fixed to the roof-rack (Figure 2.2).

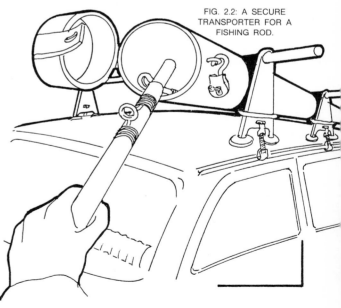

FIG. 2.2: A SECURE TRANSPORTER FOR A FISHING ROD.

Here we have the beginnings of a good idea; if it worked your dad might buy it from you, so might your friends and neighbours — the idea might even be commercial! Here you have to find out the true nature of the problem, solve it and evaluate the solution. The real creative spark comes in recognizing the problem. The principle is sound, but it needs careful consideration to make the design suit the car visually and be thief-proof.

Portable containers
Carry-cases for fishing, photography, archery, caving, climbing, bird-watching. Equipment boxes for cycle-racing, camping. Containers for gardening, to hold plants, vegetables, fruit or weeds.

Containers for the garage, for car and house maintenance, for the kitchen and bathroom (linen, dirty washing, first aid).

Display containers for museums, school exhibitions, science and art exhibitions, old peoples' homes, local history societies, nature reserves, sports club trophies.

Storage containers for a teenager's bedroom: for cosmetics, shoes, hobbies (collecting and displaying), model-making parts, telephone books, address books, notepads, writing and drawing equipment.

Leisure products
Various items of outdoor sports equipment: fishing rods, bows, skis, camping, swimming and sailing aids, cycle panniers, survival packs, first aid boxes.

Indoor sports exerciser (perhaps based on roofrack 'bungy' straps or stretchable rubber seat webbing).

New board games for the family, perhaps using electronic counting and scoring devices.

Furniture for the garden, for the patio, for listening to music, for watching television, for storing records and cassettes, for hobbies, painting, dressmaking, etc.

Domestic products
Light fittings for a variety of purposes, clocks and other timepieces, using electrical, electronic or mechanical movements, educational toys for various age groups, playthings for the home, beach, kindergarten or playschool.

Shopping trolley, household trolley for DIY equipment, car inspection crawler board.

Hand-held instruments: tweezers, tongs, servers (cutlery, corn-on-the-cob holder), small specialist tool handles, garden trowels, envelope openers, juice extractors. Paper clips, polyethylene bag fasteners, polystyrene mug holders.

Cosmetics and jewellery storage, carrying-cases for girls and women.

PROJECTS BASED ON SPECIFIC MATERIALS AND PROCESSES

In the early stages of learning how to design and make things from plastics it is worthwhile to select a project that can be made by a specific process or will exploit the properties of a specific material. For example, polythene and PVC film materials are cheap and readily available, can be welded easily and are suitable for projects ranging from kites and flying sculptural forms to inflatable furniture, fashion clothing, wallets, folders and special bags.

Sheet heat-forming, sheet bubble-blowing and press-forming of various types all lend themselves to specialized project work in which the properties of sheet materials — opaque or translucent colour or transparency, rigidity or flexibility, or varying textures — can all be exploited to suit specifications.

Glass-reinforced plastics are ideal for large, strong mouldings for chairs, playground equipment, wheelbarrows, canoes and boats. For many very small objects, resin casting, injection and simple compression moulding offer a wide variety of opportunities, and do not always require expensive machines to achieve satisfactory products.

GCSE students looking for demanding technical projects based on plastics can find a variety of interesting problems in designing and making small machines for moulding plastics. The principles are all relatively simple, and where funds are limited it can be much cheaper to make a machine than buy one. However, *safety regulations* must be followed to the letter (see Chapter 15).

Plastics are ideal for many applications but do not suit every situation. Choose a project for which your chosen material is better-suited than alternatives, such as wood, metal or concrete, and where possible allow the aesthetic qualities of the material to enhance the features of your design.

DEVICES FOR OLD AND DISABLED PEOPLE

Contact local authority welfare services for introductions to people in need. Schools and colleges can do a great deal to help individuals with special problems that cannot be catered for by commercial aids. Provided that both parties, the disabled person and the student designer, enter the arrangement without expecting too much, it is highly probable that a successful new design can be developed. Involvement in doing something worthwhile can draw out your best efforts. Teamwork, regular contact and a realistic budget and timescale are essential. And it is not enough to satisfy the strictly functional requirements: great care is needed to make the aid attractive to look at. Otherwise it will be hidden in a cupboard and, however sound it is technically, will remain unused.

PROJECTS FOR LOCAL INDUSTRIES

Project ideas can often be found in small local businesses — factories, shops, farms and horticultural centres. If you live in the country and know a local farmer, find out what his specialized agricultural work is, ask if you can watch him during his daily routine and discuss any aspects that he finds difficult, unpleasant, annoying, or dangerous. Then see if you can devise a technical aid to relieve the problem.

A garden centre may have a genuine need for special plant display equipment, temperature- and humidity-controlled plant propagators, lifting equipment, outdoor lighting, waterproof pamphlet racks. Establishing, peoples' needs and preparing a proper brief and specification are major parts of arriving at a worthwhile project. Finding out how people do things often reveals that they have never really thought about the task or set out to find an easier or better way.

DESIGNING IN PLASTICS

Basic principles

Each of the two subgroups of the plastics family has properties that affect the way the materials can be processed or used. When designing any plastics product it is necessary either to (1) select the best material available and then choose the most suitable process or (2) choose the process and then select the material. The choice of process will often be decided by the total number of products to be made, the proposed product size and shape (based on human factors, product function and aesthetic considerations) and the available processing equipment.

Designing the product is a complex procedure involving many different factors at the same time. These become design decisions which must be made before any work is started on the mould. (See the charts on pages 124 and 125.)

Moulding

What is a mould? It is a device that is used to give a plastics material *shape*. It may be hollow, in which case it is said to be a *negative* mould (Figure 2.3); or it may be convex (hump-shaped), in which case it is called a *positive* mould (Figure 2.4). The plastics article made in or on a mould is called a moulding. The surface of the moulding is usually a mirror image of the mould, — that is, the opposite way round. The designer of a plastics product has to think about the positive and negative shapes of both the mould and the moulding at the same time.

In some processes it is possible to use a simple mould, which can be either positive or negative, the choice being made on the basis of which surface is most important in the finished product. There are, however, several processes where the mould is in two halves, one positive, the other negative (Figure 2.5). One fits inside the other, leaving a space called a *cavity*, which is filled with plastics material. The plastics material is forced into the cavity, hardens into a solid, is removed and has good surfaces on *both* sides. Injection moulding is a typical process of this type, and is a method of *pressure casting*. Often it is impractical to modify a mould and it is unusual for mouldings to be altered once they have been formed.

In some processes a mould can be referred to as a *plug*, a *former*, a *preformer*, a *jig* or a *tool*. In industry a simple mould can cost as little as a few hundred pounds; a complex one can cost many thousands of pounds. A small part of the cost of a mould is paid for each time someone buys a moulding. Usually thousands of mouldings are made from a mould.

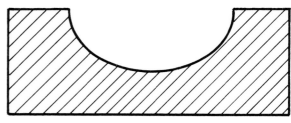

FIG. 2.3: CONCAVE (NEGATIVE) MOULD.

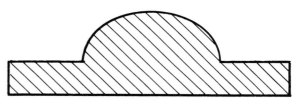

FIG. 2.4: CONVEX (POSITIVE) MOULD.

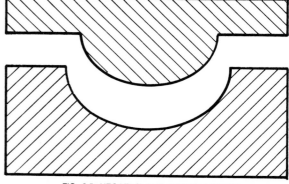

FIG. 2.5: NEGATIVE AND POSITIVE MOULD.

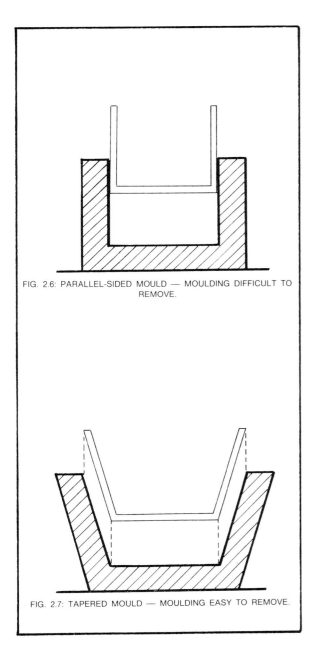

FIG. 2.6: PARALLEL-SIDED MOULD — MOULDING DIFFICULT TO REMOVE.

FIG. 2.7: TAPERED MOULD — MOULDING EASY TO REMOVE.

Rule 1: Taper and draft angle

Most processes require moulds and mouldings to be tapered so that they can be withdrawn from each other. Several names are given to this: *taper*, *withdrawal angle*, *angle of draw* and *draft*. Figure 2.6 shows a parallel-sided vessel and Figure 2.7 a tapered vessel, both being withdrawn from moulds. The parallel-sided moulding is extremely difficult to remove because the walls remain in contact during separation. On the other hand, the tapered vessel separates completely from the mould immediately it is lifted.

There are occasions, of course, when it is essential to have parallel sides, and then one has to accept the difficulties of removing the moulding from the mould and the extra time it takes. As a design principle however, always aim to provide some taper if at all possible (Figure 2.8). When two identical mouldings are going to meet face to face (suitcase halves, for example), make the top part of the mould parallel-sided and taper the lower part. This also adds considerable strength to the sides of the moulding. (See Figures 2.11 and 2.14.)

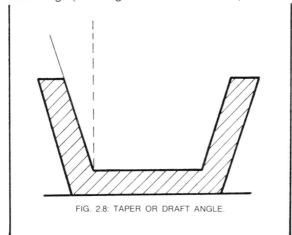

FIG. 2.8: TAPER OR DRAFT ANGLE.

Rule 2: Corner radii

The next important design feature, again applicable to most processes, is to design a *radius* whenever a corner is necessary (Figure 2.9). A radius allows the material to 'turn a corner' slowly, thus maintaining a constant section (thickness). A sharp, angular corner, on the other hand, is liable to be thicker in section, which can cause it to be brittle in some materials, and generally difficult to clean and handle. Always aim to provide the largest radius (gentlest curve) possible, in proportion with the rest of the product form. (Note: further information on corner design is given in other sections throughout the text — for example, the sections on injection moulding with reference to corner and rib design (page 62) and on glass-reinforced plastics mouldings (page 95).

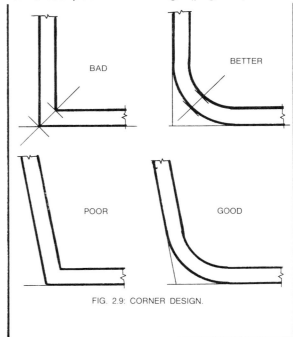

BAD

BETTER

POOR

GOOD

FIG. 2.9: CORNER DESIGN.

21

Rule 3: Surfaces

The next important general design feature is to mould-in *curvature* to all surfaces where possible, avoiding large flat areas. Rounded shapes are stronger and less likely to become distorted. Flat shapes tend to warp and cause adjacent side walls to bow inwards (Figure 2.10). When flat shapes are essential, moulded-in ribbing or other reinforcement should be considered (Figures 2.11 and 2.12). Hollow (concave) and bulbous (convex) curvatures overcome most of the distortion effects, add visual and physical strength and often mean that thinner material can be used. Figures 2.11 to 2.14 show the preferred design treatment of relatively large surface areas.

Vacuum formed seed container — the moulded shape gives added strength to very thin material.

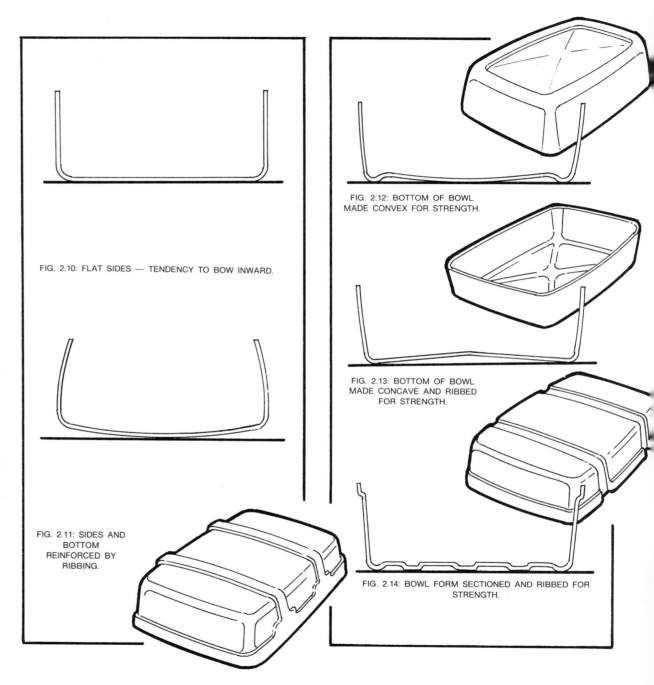

FIG. 2.10: FLAT SIDES — TENDENCY TO BOW INWARD.

FIG. 2.11: SIDES AND BOTTOM REINFORCED BY RIBBING.

FIG. 2.12: BOTTOM OF BOWL MADE CONVEX FOR STRENGTH.

FIG. 2.13: BOTTOM OF BOWL MADE CONCAVE AND RIBBED FOR STRENGTH.

FIG. 2.14: BOWL FORM SECTIONED AND RIBBED FOR STRENGTH.

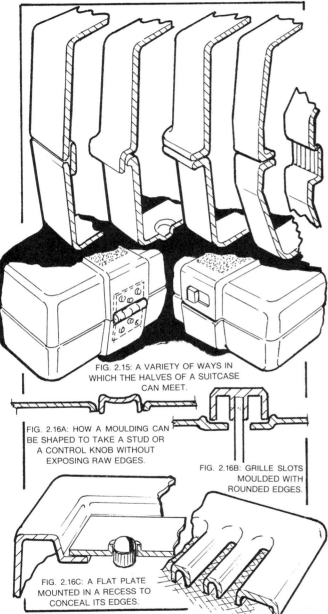

FIG. 2.15: A VARIETY OF WAYS IN WHICH THE HALVES OF A SUITCASE CAN MEET.

FIG. 2.16A: HOW A MOULDING CAN BE SHAPED TO TAKE A STUD OR A CONTROL KNOB WITHOUT EXPOSING RAW EDGES.

FIG. 2.16B: GRILLE SLOTS MOULDED WITH ROUNDED EDGES.

FIG. 2.16C: A FLAT PLATE MOUNTED IN A RECESS TO CONCEAL ITS EDGES.

Rule 4: Detail design features

When you work in plastics you need to prepare designs more fully than with other materials, because moulds have to be built before the plastics materials can be formed. In planning the product and the mould it is wise to consider the detail design features at the same time as the overall shape. For example, it is better to mould the mounting points for suitcase locks, catches, and hinges into the form (Figure 2.15), rather than juggle them into position on the finished mouldings. If control panels, knobs, buttons, or grilles have to be let into a surface (Figure 2.16), it is better to turn the edges of the moulding in so that raw edges are hidden, which gives a more professional finish (See page 96).

Colour, contrasts with accessories made of non-plastics materials, and the surface finish of the mouldings are all important aesthetic factors that have to be thought out in the early stages. The sizes of mouldings in relation to each other, the way they are assembled, access points for fitting mechanisms and maintenance — all these are important features.

All plastics materials have good electrical insulating properties, but anyone designing a housing for electronic components should remember that sometimes metal shielding will be necessary to exclude magnetic and electrical interference.

Most rigid plastics materials are self-supporting, especially when they are given rounded shapes. However, many materials will *creep* (become permanently deformed) or at least become temporarily distorted if they are used in large surfaces and left unsupported. Thermoplastics are most likely to suffer in this way, especially when they are left in warm environments, for example, near radiators or in a car on a hot day.

Deformation (change in shape) can be caused by leaving the object in the standing position, by internal stress from moulding processes or by heavy objects resting against the moulding. A thermoset that is regarded as a true structural material is glass-reinforced polyester resin, in which the strength comes from the glass reinforcement.

The foremost processes for making plastics products are moulding operations, rather than machining. The design principles discussed here have therefore centred on good moulding practice. When components have to be machined from cast shapes, or cut from thick sheet, then all corners should be radiused. Sharp corners must be avoided because cracks will start at these points and 'run' quickly through the material.

Hinge details on polystyrene injection moulded boxes.

23

The photograph shows a hairbrush made in two components from two different materials. The designer has taken advantage of the rigid properties of polystyrene to create the strong handle and brush back. Advantage has also been taken of the flexible properties of polyethylene to make the bristle section. This was originally moulded with all the bristles parallel to one another and then slightly curved during assembly.

Two sizes of durabeam torch. Close examination of these torches show good design detailing. The overall appearance of the product is exciting, novel and practical.

PROJECTS AND ACTIVITIES

(1) Three dimensional maps can be a most useful teaching aid. They can be made cheaply by vacuum forming thin sheet plastics over a positive mould. Discuss with your Geography Teacher the design of a route for a river from small streams in some mountains to the sea and then design and make a positive mould from either clay or layers of thick card.

(2) Biology field study trips are run annually by many schools. Find out the conditions and problems for catching insects and small mammals. If possible identify one species of insect, for example mosquitos, or one habitat, perhaps insects that fly between one and three metres above the ground, and then design a suitable net or trap. Plan your work to include research, analysis, idea development, modelling, testing and evaluation. Prototype your design if possible.

(3) The same biology field trips often require delicate scientific instruments such as microscopes. Is it possible to improve the carrying case for a microscope e.g. make it lighter, easier to carry or easier to use out in the open?

(4) You will often see your school caretaker carrying equipment round the school and working at a wide variety of jobs. Does he or she need any form of special trolley to make these tasks easier? Find out — it could be the source of at least one very interesting project.

(5) Night classes are run in many schools in the evenings during the winter months. In large schools the buildings are often far apart making it difficult for night school visitors to find their way in the dark. Investigate the need for an illuminated sign system or outdoor lighting system to show people the way. Club together with three friends and as a group project, design and make some signs for exterior use.

(6) String, percussion and wind musical instruments based on those made by primitive peoples can all be made in plastics materials. The way sounds are made can be studied both scientifically and experimentally through models. Research one type of instrument and then design and make your own version. Experiment with plastics bottles and margarine tubs.

(7) Measuring and scoring devices are often needed for athletics but are very expensive. Ask your PE teacher if he or she needs any special equipment of this type and the methods used at present. See if you can design a better more accurate device or system.

(8) Imagine you have a friend who is blind and unable to read my notes on pages 21–23. Using a tape recorder and your own words describe to your friend all the points mentioned. If you do not have a tape recorder then find a way of making models or using products to describe each point but do not use any words. If one half of the class makes models and the other tape recordings, you can then judge each others' work blindfolded.

(9) Find a plastic pack or bottle that is no longer needed. Cut it in half and examine it closely making notes about any detail features. Then use this book to see if you can describe how it was made and the shape of the mould that would have been used in its manufacture. Illustrate your answer and be prepared to give a four minute talk to the class about it.

3 SHAPING SHEET PLASTICS USING HEAT

SHEET MOULDING

Sheet thermoplastics change shape easily when heated. The following pages outline the procedures for preparing these materials and then show a wide variety of methods that can be used to achieve basic moulded shapes.

The methods and designs shown are *not* to be copied but used as the basis for making mouldings for your own products. The detailed illustrations and descriptions should provide you with enough information to design your own products and moulds. Be prepared to experiment and develop your own methods to suit your particular needs. In most cases if you make a mistake when forming thermoplastics materials it is possible to reheat the sheet and to try again. This is one of the valuable attributes of thermoplastics.

It is sensible to keep a box of off-cuts. These are ideal to experiment with and often come in useful — especially for small pieces of jewellery.

Warning: When handling hot plastics sheets always wear protective leather gloves. Never allow materials to burn and do not leave them on heaters or in an oven unattended.

CUTTING SHEET PLASTICS

Figure 3.1 shows a sheet of paper-covered acrylic being marked out with a pencil. The protective paper should be kept on the plastics material for as long as possible, especially during sawing and drilling. Sawdust can scratch the glossy surface.

Straight cuts may be made on the sheet by deep scoring of both sides (Figure 3.2). The material should be carefully broken along the score line. It is necessary to hold both parts of the sheet, supporting one side so that the score

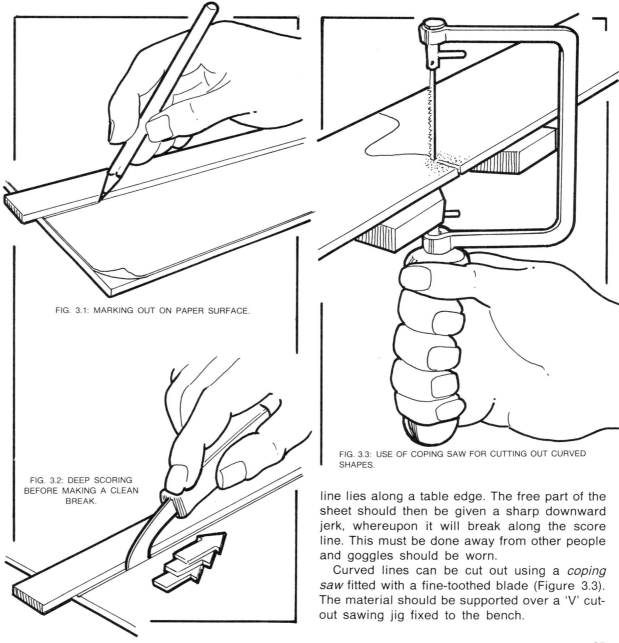

FIG. 3.1: MARKING OUT ON PAPER SURFACE.

FIG. 3.2: DEEP SCORING BEFORE MAKING A CLEAN BREAK.

FIG. 3.3: USE OF COPING SAW FOR CUTTING OUT CURVED SHAPES.

line lies along a table edge. The free part of the sheet should then be given a sharp downward jerk, whereupon it will break along the score line. This must be done away from other people and goggles should be worn.

Curved lines can be cut out using a *coping saw* fitted with a fine-toothed blade (Figure 3.3). The material should be supported over a 'V' cut-out sawing jig fixed to the bench.

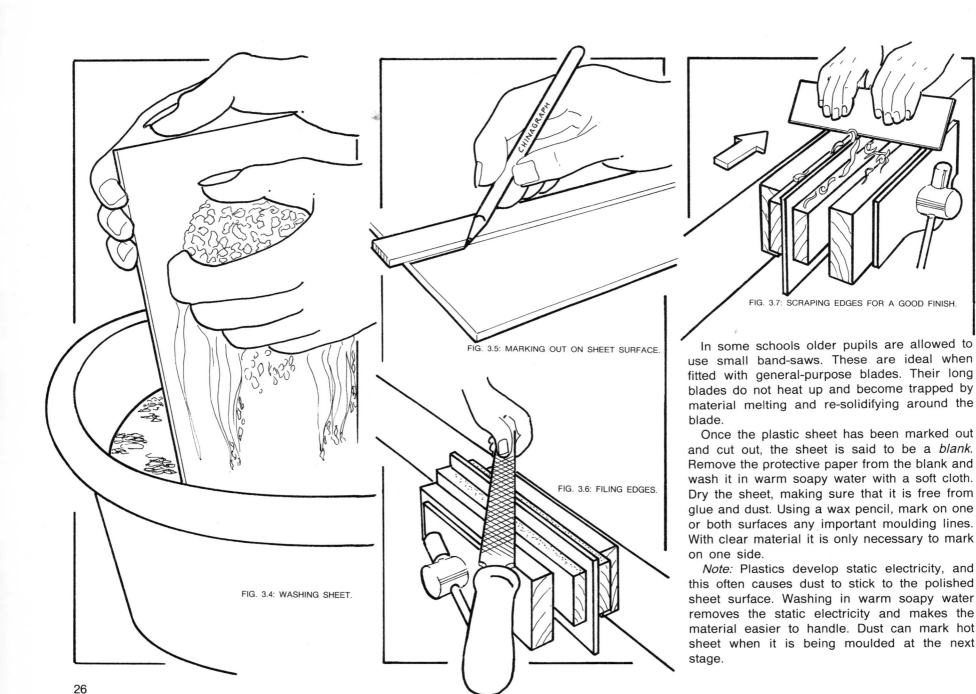

FIG. 3.5: MARKING OUT ON SHEET SURFACE.

FIG. 3.7: SCRAPING EDGES FOR A GOOD FINISH.

FIG. 3.6: FILING EDGES.

FIG. 3.4: WASHING SHEET.

In some schools older pupils are allowed to use small band-saws. These are ideal when fitted with general-purpose blades. Their long blades do not heat up and become trapped by material melting and re-solidifying around the blade.

Once the plastic sheet has been marked out and cut out, the sheet is said to be a *blank*. Remove the protective paper from the blank and wash it in warm soapy water with a soft cloth. Dry the sheet, making sure that it is free from glue and dust. Using a wax pencil, mark on one or both surfaces any important moulding lines. With clear material it is only necessary to mark on one side.

Note: Plastics develop static electricity, and this often causes dust to stick to the polished sheet surface. Washing in warm soapy water removes the static electricity and makes the material easier to handle. Dust can mark hot sheet when it is being moulded at the next stage.

THERMOPLASTIC PROCESSES USING SHEET MATERIALS

Strip bending

This process is also suitable for most other thermoplastic materials.

Warning: Watch the material all the time as it is being heated and do not let it burn, blister or catch fire. Do not hold your fingers over the strip heater element or touch the hot surrounding area.

Place the material on the strip heater rests as shown in Figure 3.8, so that the bend line lies directly above the heater element. The height setting for the rests will affect the area of sheet that is heated and determine the size of the radius. After *every 10 seconds* turn the sheet over and heat the reverse side along the same bend line. It is necessary to keep turning the plastics material during the heating process because the sheet may remain cold on one side while the other side is extremely hot. By turning the sheet every few seconds the heat moves through the material from both sides, preventing both surfaces from becoming too hot. This becomes more important as thicker materials are used. With thick material the heater setting should be low and the total heating period longer.

Keep turning the sheet until it becomes soft and floppy along the bend line, and when the material is soft, place it on the jig and hold it firmly until the softened area becomes rigid again (Figure 3.10). The bend should have the same radius all the way along its length: if it varies from one end to the other, either you have allowed the material to wander away from the jig while it cooled, or it has become overhot in the centre (Figure 3.9).

Material that has not formed properly can often be reheated and reformed. Two tips: (1) A line marked on the jig will often help to locate the edge of the sheet in relation to where the bend occurs. (2) If several bends are to be worked into the same sheet, work out which should be done first. Otherwise it may be difficult to heat the sheet because the early bends get in the way. Fold a paper shape to see which order to make the bends in.

A strip heater, sometimes known as a line heater, can be used to mould many complicated shapes. Most thermoplastic materials can be used on this piece of equipment, but remember that you may need more, or less, time for heating, according to the thickness and chemical composition of the sheet. Pressure marks or marks from the mould will be noticeable on both sides of very thin material.

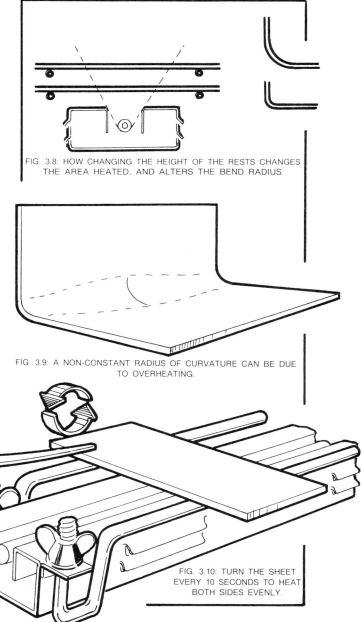

FIG. 3.8: HOW CHANGING THE HEIGHT OF THE RESTS CHANGES THE AREA HEATED, AND ALTERS THE BEND RADIUS

FIG. 3.9: A NON-CONSTANT RADIUS OF CURVATURE CAN BE DUE TO OVERHEATING.

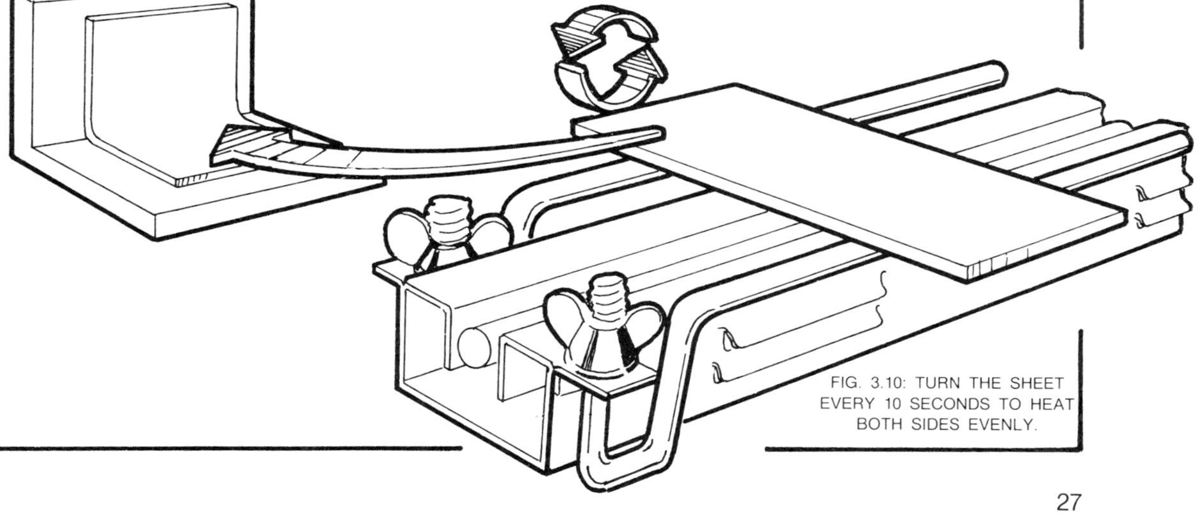

FIG. 3.10: TURN THE SHEET EVERY 10 SECONDS TO HEAT BOTH SIDES EVENLY.

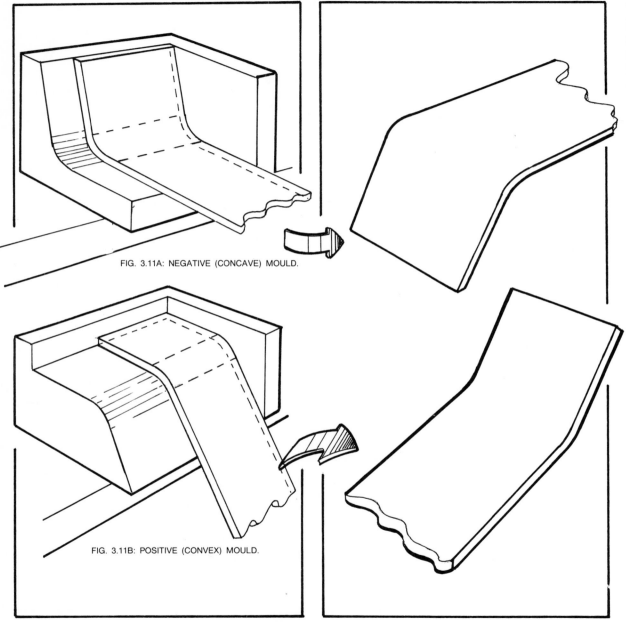

FIG. 3.11A: NEGATIVE (CONCAVE) MOULD.

FIG. 3.11B: POSITIVE (CONVEX) MOULD.

The jig or former

It is useful, although not essential, to have a jig so that when the material has been heated it may be bent accurately to shape. The jig may be a negative shape *into* which the material may be placed (Figure 3.11A) or it may be a positive shape *onto* which it may be pressed (Figure 3.11B). A narrow band of plastic will be hot and floppy after heating, and it is wise to use a jig to make sure that the angle of bend is correct and that the material is square. The radius of bend will also be controlled on a jig, but may alter if the material is left unsupported. Jigs are used to ensure repeatable dimensional control.

A jig can be made from a solid piece of wood, planed and radiused to the desired shape, or it may be made from two or more pieces glued and screwed together. A good surface is important on the jig to prevent it from making pressure marks.

Before making a jig or using any that are available in the workshop, decide which surface of the plastics material will be most important. This is the one that should *not* be placed in contact with the jig. Then make or use the jig that will give the correct radius bend on the less important surface, so pressure marks are less likely to show. When the jig is ready, place it near the heater.

Remove the protective paper from the acrylic sheet, wash it in warm soapy water and mark the bend line on both sides with a wax pencil.

Once it has been heated, the material must be placed on the jig *quickly*.

Ovenwork

The electric oven is an essential piece of equipment for many plastics sheet processes. Most ovens can be set to specific temperatures; when this temperature is reached an electrical control (a *thermostat*) will switch the electricity on and off so that the temperature stays the same for as long as is necessary. Some ovens contain a fan to keep the air moving so that all parts of the oven are at the same temperature.

Different types of plastics materials soften at different temperatures. Materials of the same type but of different thicknesses will soften at the same temperature, but the thicker they are, the longer they will take to heat up to that temperature. Remember that plastics are poor conductors of heat.

When you heat any very thick material (over 6 mm), it is best to heat the material through thoroughly at a few degrees below the moulding temperature and then increase the temperature just before moulding. This prevents surface blistering, which can occur when a material is heated for a long time at the higher moulding temperature.

Acrylic sheet

Most acrylic sheets are made by a casting process that is very carefully temperature-controlled. Made in this way, it is free from stress — that is, there are no tension forces in the material.

When such a sheet is heated to moulding temperature (165–170°C), it keeps its shape and

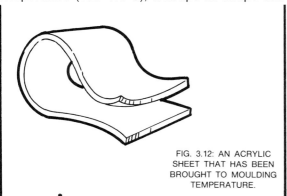

FIG. 3.12: AN ACRYLIC SHEET THAT HAS BEEN BROUGHT TO MOULDING TEMPERATURE.

thickness although it is relaxed and floppy. At this temperature the sheet may be formed on a mould and allowed to cool. Several moulding operations and mould types are explained in the following pages, but it must be remembered that some force must be applied to the sheet to make it take a new shape and that this force will put stresses into the material. As the moulding cools, these stresses will be 'locked' in the material. If the moulding is unsatisfactory at the first attempt, it may be returned to the oven and reheated. The material will return to its original flat state, become the original thickness and shape, and in its floppy condition can be re-moulded. When floppy it is free from internal stresses.

To test whether the material has reached moulding temperature, bend a corner over and touch it against the sheet. If it bends over easily and feels 'tacky', the sheet is ready for moulding (Figure 3.12).

Warning: Remember to handle materials in the oven only when wearing heat-resistant leather gloves.

A deep moulding must be made in one operation: the sheet cannot be partly formed, reheated and made deeper in a second operation.

Products can be 'stress-relieved' through a form of annealing (see pages 117–118).

Figure 3.13 shows how a large radius is put into a strip of hot material. The mould has been cut out from a solid slab of timber and has a spacer, which acts as a stop-rest for the sheet, and a rubber strip to hold the material against the mould. The hot plastics sheet is taken from the oven and one end is placed against the spacer. The rest of the material is gently laid over the curved shape of the jig, followed quickly by the rubber, which is drawn firmly over the

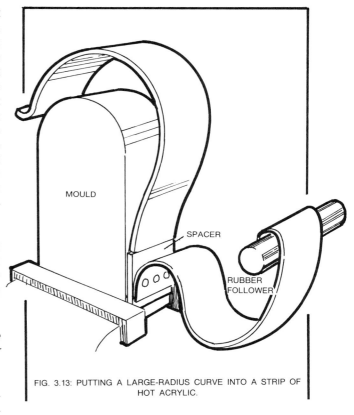

FIG. 3.13: PUTTING A LARGE-RADIUS CURVE INTO A STRIP OF HOT ACRYLIC.

plastics material to make it take the shape of the jig.

A variety of formers for producing more complicated shapes is shown in Figures 3.14–18. A wavy curve is easily produced in a strip of thermoplastic sheet (Figure 3.14). It will be noticed that both the negative (A) and positive (B) pieces of this matched mould have been cut from the same slab of material. The hot acrylic sheet only needs to be placed on the positive former (B), and negative former (A) then pressed firmly down onto it, to make it follow the shape.

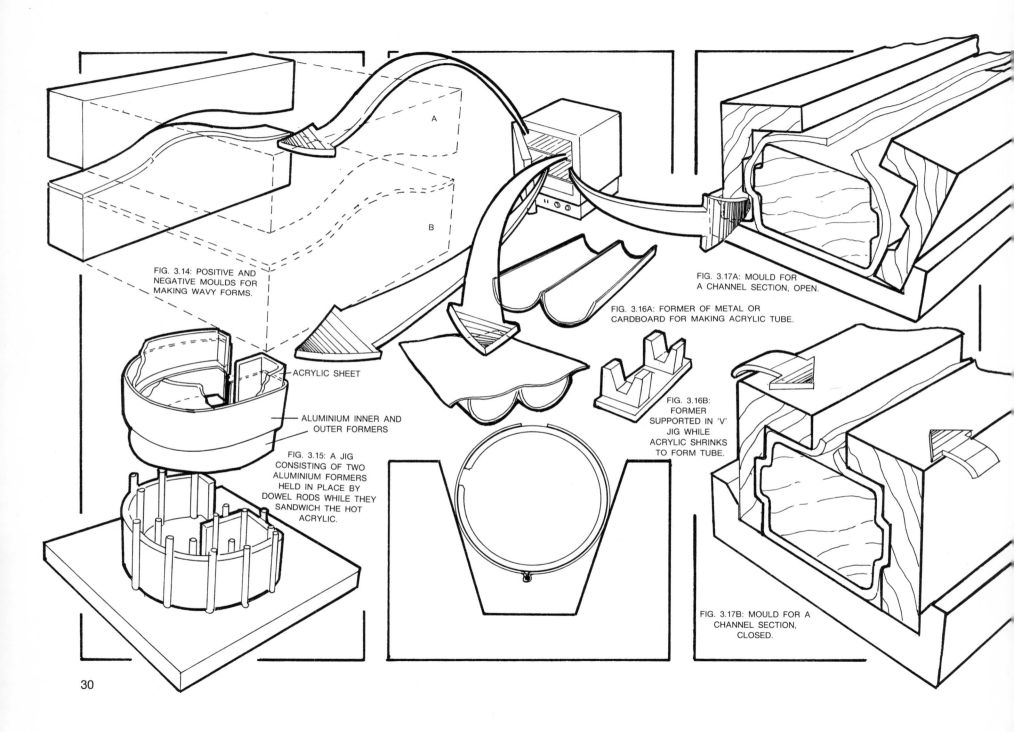

FIG. 3.14: POSITIVE AND NEGATIVE MOULDS FOR MAKING WAVY FORMS.

ACRYLIC SHEET

ALUMINIUM INNER AND OUTER FORMERS

FIG. 3.15: A JIG CONSISTING OF TWO ALUMINIUM FORMERS HELD IN PLACE BY DOWEL RODS WHILE THEY SANDWICH THE HOT ACRYLIC.

FIG. 3.17A: MOULD FOR A CHANNEL SECTION, OPEN.

FIG. 3.16A: FORMER OF METAL OR CARDBOARD FOR MAKING ACRYLIC TUBE.

FIG. 3.16B: FORMER SUPPORTED IN 'V' JIG WHILE ACRYLIC SHRINKS TO FORM TUBE.

FIG. 3.17B: MOULD FOR A CHANNEL SECTION, CLOSED.

When you make matched moulds, you must allow for the thickness of the plastics material that you will finally use. This is very important

This product is designed to clip onto the arm of a garden chair and is intended to hold a drinks beaker, book/newspaper and other small items. It was made from sheet foamex PVC heated in an oven and formed between the sheet metal jig. When cold the shaped material was cemented, the radii in the corners filled and then sprayed with car aerosol paint.

when you have to mould sharp, deep shapes.

Figure 3.15 shows two aluminium sheets used to support a strip of hot acrylic sheet while it cools. The inner and outer sheets of shaped metal are supported by dowel pegs.

Figure 3.16 shows a former for making a tube. Clear acrylic tube and household pipe made from various thermoplastics materials are usually the only kinds of plastic tubes available to schools for project work. It is useful to know how to make acrylic tube of any diameter. The length and diameter are then limited only by oven size. Hinge two metal or strong cardboard channel sections together with adhesive tape (Figure 3.16A). Heat an accurately cut sheet acrylic blank and lay it over the mould, which is then gently closed, and laid into a 'V' jig (Figure 3.16B). The hot acrylic sheet shrinks as it cools. Do not force the mould shut as that would cause distortion in the tube.

Note that it is not possible to make an accurately shaped thermoplastic tube by wrapping hot sheet round a drum. It always results in two flats occurring at the point where the sheet ends meet.

A mould for producing a 'channel' section is shown in Figure 3.17A, with hot material in position. In Figure 3.17B the mould is being closed onto the material to create side shaping.

Figure 3.18A shows a peg and press mould, to make 'wavy edge' bowls and dishes. A flat sheet of 9 mm or 12 mm plywood is drilled to receive short lengths of dowel rod. The length of these dowel rods and their spacing depends on the proportions of the finished article. The press disc, which may be made from the same plywood, or from any rigid material, should fit comfortably inside the ring of dowel rods. A press disc that is a relatively close fit will give steep sides (Figure 3.18B) while a much smaller press disc will give very shallow edges (Figure 3.18C).

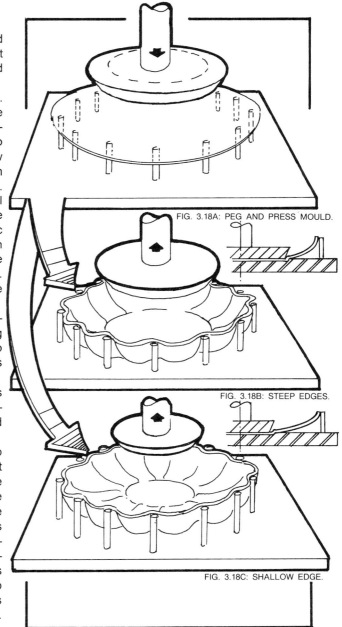

FIG. 3.18A: PEG AND PRESS MOULD.

FIG. 3.18B: STEEP EDGES.

FIG. 3.18C: SHALLOW EDGE.

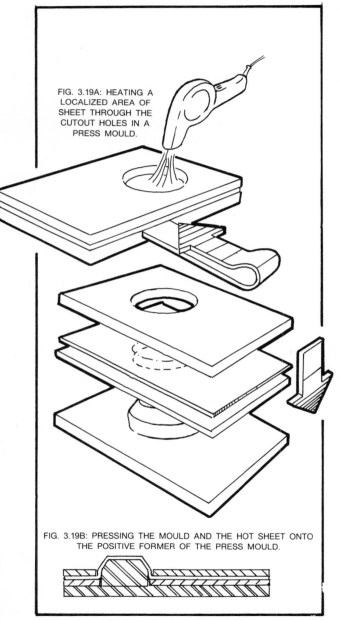

FIG. 3.19A: HEATING A LOCALIZED AREA OF SHEET THROUGH THE CUTOUT HOLES IN A PRESS MOULD.

FIG. 3.19B: PRESSING THE MOULD AND THE HOT SHEET ONTO THE POSITIVE FORMER OF THE PRESS MOULD.

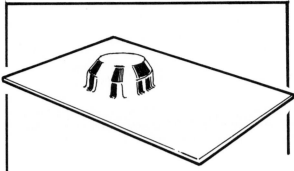

Localized shaping of thermoplastics sheet material

Sometimes you will want to heat a localized area of thermoplastics sheet to mould a shape. For example, you may wish to press a shallow circular recess into it. For this you would need a hair drier or other hot-air blower and a press mould (a board with a positive form and two boards with holes cut out).

The upper board with the hole may be used as a mask during the heating operation (Figure 3.19A), as it is only this area that will be moulded. The sheet should be turned over periodically, keeping both holes and the heated area exactly aligned.

When the localized area has become fully softened, place the board with the boss on the table and the thermoplastic sheet, sandwiched between the cutout boards, on top. Then press firmly down (see Figure 3.19B). The boss should be smaller all round by the thickness of the plastics material, so that it can press easily into it, creating the required shape. It may be necessary to raise the board with the cutout off the table if the boss is deeper than the board thickness.

Whenever possible, avoid designing a shape too near to the edge of the sheet. When this occurs, use a larger plastics blank than is really necessary, and trim it back after the moulding operation.

Always allow sufficient flat material around the moulded area to resist distortion: a breadth equal to the radius of the boss is ideal. Note that the height of the boss must be less than its radius.

Plastics tube can often be bent by localised heating using a strip bender, a hair dryer or a hot air welder (Figure 3.20).

The tube must be rotated constantly during the operation and when it is hot and soft moved to a supporting jig and allowed to cool at the required angle. During heating with hot air equipment a curved sheet metal reflector plate will direct the hot air blast around the tube to help bring it all rapidly up to forming temperature.

It is most likely that the tube will have originally been produced by extrusion and if overheated will shrink slightly and bubble. Different plastics materials react in different ways so always experiment first.

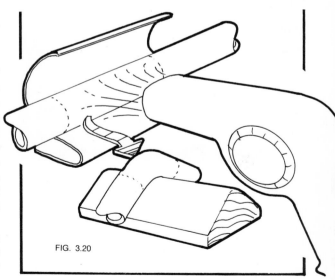

FIG. 3.20

Plastic 'memory'

All thermoplastics have 'memories' and, after being stressed, will return on heating to their original form. Acrylics are particularly good for demonstrating this because they are originally cast flat and so are free from stress. Materials like polystyrene, ABS, polyethylene, and polypropylene, on the other hand, are produced by extrusion and calendering processes (see Chapter 7, page 66) which put stress into the material. They roll up when heated in an oven and are difficult to control as they return to their unstressed shape.

Here is a project that makes use of the 'memory' of an acrylic block.

Heat the block (which should be at least 6 mm thick) slowly to 170°C for 15 minutes (Figure 3.21). Place a wire pattern onto the hot acrylic block and sandwich them between steel plates (Figure 3.22). Close the assembly firmly in a vice and allow the assembly to cool (Figure 3.23).

When the assembly is cold, remove it from the vice and remove the wire from the block (Figure 3.24). File or machine the block flat, down to the lowest part of the impression (Figure 3.25). Polish the surface using fine wet-and-dry paper (600/1200 grade) and water, and then polish and buff.

Reheat it in the oven to 170°C for 15 minutes (Figure 3.26). Remove it, allow it to cool, and trim it to shape (Figure 3.27).

When the acrylic material is heated and moulded in this way, the wire pattern squashes the acrylic sideways. Removing the material surrounding the impression leaves the material under the pattern unaffected. It retains a 'memory' of its original thickness. Reheating allows the pattern to return to its original height and the rest of the block to shrink to its former area.

This process can be used for making

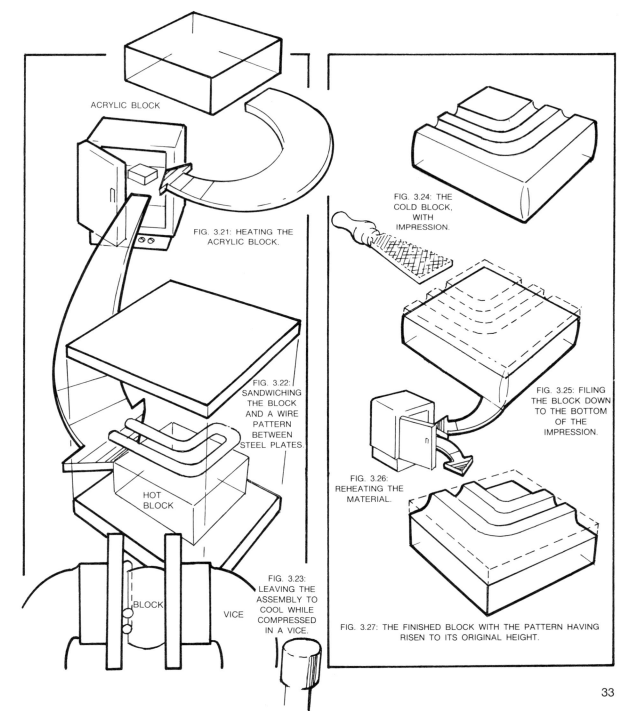

ACRYLIC BLOCK

FIG. 3.21: HEATING THE ACRYLIC BLOCK.

FIG. 3.22: SANDWICHING THE BLOCK AND A WIRE PATTERN BETWEEN STEEL PLATES.

HOT BLOCK

BLOCK

VICE

FIG. 3.23: LEAVING THE ASSEMBLY TO COOL WHILE COMPRESSED IN A VICE.

FIG. 3.24: THE COLD BLOCK, WITH IMPRESSION.

FIG. 3.25: FILING THE BLOCK DOWN TO THE BOTTOM OF THE IMPRESSION.

FIG. 3.26: REHEATING THE MATERIAL.

FIG. 3.27: THE FINISHED BLOCK WITH THE PATTERN HAVING RISEN TO ITS ORIGINAL HEIGHT.

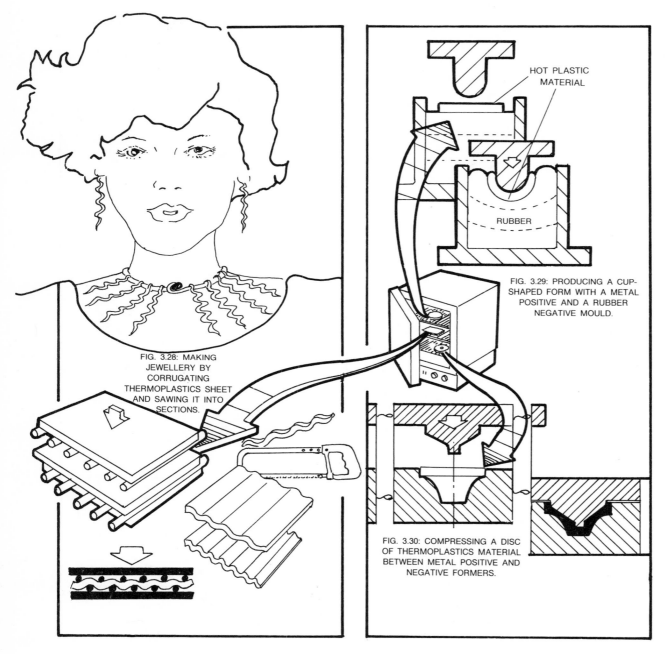

FIG. 3.28: MAKING JEWELLERY BY CORRUGATING THERMOPLASTICS SHEET AND SAWING IT INTO SECTIONS.

HOT PLASTIC MATERIAL

RUBBER

FIG. 3.29: PRODUCING A CUP-SHAPED FORM WITH A METAL POSITIVE AND A RUBBER NEGATIVE MOULD.

FIG. 3.30: COMPRESSING A DISC OF THERMOPLASTICS MATERIAL BETWEEN METAL POSITIVE AND NEGATIVE FORMERS.

jewellery and the lids to small decorative boxes. The rest of the box can be made from round or square tube.

Three methods for press-forming thermoplastic sheet using 'matched' split moulds

These methods use the principle that hot thermoplastic block or sheet will flow under pressure in a closed mould.

1: Making jewellery by press-forming hot acrylic sheet between two steel moulds. These are made from rods brazed to plates (Figure 3.28) and produce corrugated sheet. This is then sawn into thin 'snakes', polished and drilled. This is also possible with laminated PVC and CAB sheet of different thicknesses. Experiment!

2: Figure 3.29 shows a three-dimensional press-forming using a steel positive and rubber negative mould. The rubber is contained in a steel tube welded to a steel base. A small fly press or book press is ideal for pressing shapes by this method.

3: Using a mould made in two metal halves, with a space for the plastics material (Figure 3.30). The halves are aligned using guide pins. The block of plastics material must have a volume as nearly equal to that of the mould cavity as possible, and a shape that allows it to be fitted while hot into the negative cavity of the mould.

The mould halves are then closed and the pressure increased until the mould is fully closed, and a small amount of excess material begins to ooze from the cavity. When the material has cooled sufficiently but is still warm, the pressure is relaxed and the mould opened.

Press-forming of thermoplastic sheet materials

This process is suitable for making small products like the boat hull shown here, or shallow box shapes, or small positive mouldings.

The positive mould is a strong solid form, slightly tapering upwards from the base, and deeper than the finished moulding (Figure 3.31A). There should be no recesses or hollow shapes, and, in profile, angles should not change sharply, leaving concave 'corners'. The size of the mould is governed by the size of the sheet acrylic blank that can be heated in the oven. The thermoplastic blank must be larger than the distance across the mould from base to base — that is, including the height of the side walls. The mould height should not be greater than half the width.

For small mouldings, 150 mm square or less, the mould will require only one press ring. The press ring is used to force the hot plastics sheet down over the positive former and it must therefore be rigid and strong, 9 mm plywood with a surround of 50 mm is usually strong enough. It should follow exactly the ground plan of the base of the positive mould, but it must be cut to give an all-round clearance equivalent to the thickness of the thermoplastic sheet to be used.

For larger mouldings it is necessary to use two press rings. One is cut with the clearance for the thickness of the plastics material, while the other should be an accurate fit. In use the two rings are exactly aligned with each other, with the hot thermoplastic material lightly gripped between the two. This sandwich assembly is then forced down over the positive mould, ring 2 acting as a locator or guide (Figure 3.31B).

This stretches the hot plastics sheet and at the same time allows it to slip a little as necessary until the complete assembly has reached the base of the mould. The material is allowed to cool and harden for a few minutes. When the plastics material has become rigid, but while it is still hot, the moulding is removed from the mould. On large mouldings a good deal of shrinkage takes place as cooling occurs. For mouldings that are to match with other components made in different materials, it is necessary to experiment to find out the amount of shrinkage that takes place on cooling; otherwise the finished plastics component may be too small.

When you design mouldings that are to be made by this process, always provide the largest possible radii and reasonable angles of taper — at least 3° on each side.

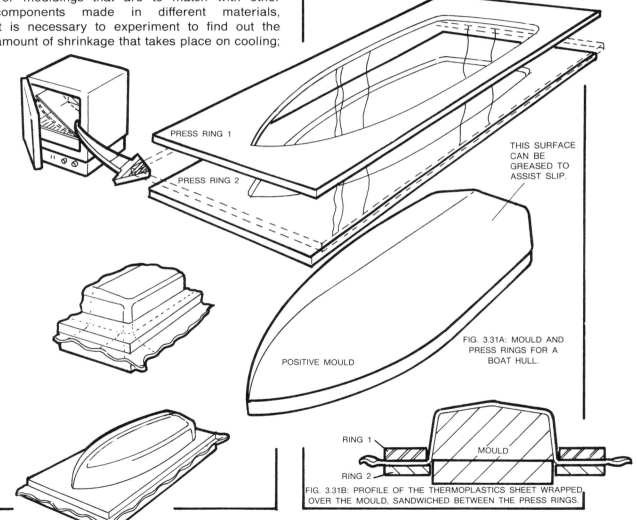

PRESS RING 1

PRESS RING 2

THIS SURFACE CAN BE GREASED TO ASSIST SLIP.

POSITIVE MOULD

FIG. 3.31A: MOULD AND PRESS RINGS FOR A BOAT HULL.

RING 1

RING 2

MOULD

FIG. 3.31B: PROFILE OF THE THERMOPLASTICS SHEET WRAPPED OVER THE MOULD, SANDWICHED BETWEEN THE PRESS RINGS.

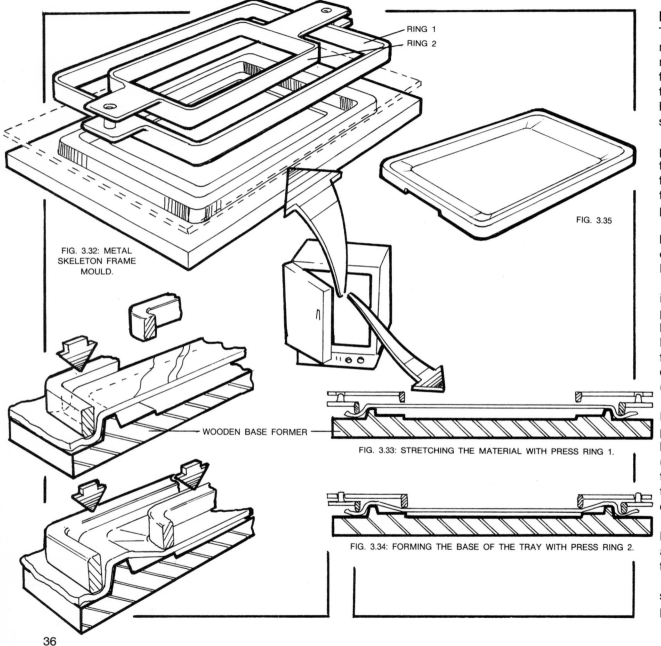

RING 1
RING 2

FIG. 3.32: METAL
SKELETON FRAME
MOULD.

WOODEN BASE FORMER

FIG. 3.33: STRETCHING THE MATERIAL WITH PRESS RING 1.

FIG. 3.34: FORMING THE BASE OF THE TRAY WITH PRESS RING 2.

FIG. 3.35

Metal skeleton frame moulds

There are many occasions when simple plastics mouldings are required but must be free from mould marks on both surfaces. Steel skeleton frame moulds are often the only means by which these can be made from sheet materials. Figure 3.32 shows a mould for a tray using two steel press rings and a wooden former.

Ring 1 acts as a clamping ring to hold the hot plastics sheet in position over the wooden base former. This ring is larger than the wooden former by an amount equal to the thickness of the plastic to be moulded. Ring 1 is fitted with pegs to locate ring 2 correctly.

Ring 2 is of the right size to create the interior bottom shape of the tray. It is designed with an extended tab at each end, drilled to fit over the location pegs on ring 1.

Inside the wooden former a hardboard spacer is fitted. This prevents the main area of the plastics sheet from 'bottoming' on the base-board. When the mould system is used, the hot plastics sheet is laid over the wooden base former. Ring 1 is pressed firmly down and clamped, and the material stretches taut (Figure 3.33).

Ring 2 is then pressed into the hot sheet, locating with the guide pins on ring 1. As it presses down, the hot sheet is lightly gripped between ring 2 and the hardboard spacer (Figure 3.34). Ring 2 is then firmly clamped until the sheet has become rigid again. As soon as the material is firm, but while it is still hot, the clamps are released and the moulding removed.

When it has been trimmed, a small tray has been made (Figure 3.35). The only visible marks are at the points where the material touched the frames or the base mould.

Skeleton frame moulds may be made from steel rod and tube, brazed together to create the positive corners of the mould (Figure 3.36).

The mould itself is a positive form whose appearance is only outlined by the steel skeleton. When a vacuum box is used, the entire shape has to be mounted upside-down on either a large threaded plunger or a compressed air cylinder; the complete assembly is mounted on a frame placed centrally over a vacuum box.

The vacuum box (Figure 3.37) may be made from heavy sheet plywood lined with GRP (see page 93). The box has interchangeable top platforms, each of which must be able to rest on a surrounding flange covered with neoprene foam rubber. Each interchangeable platform has an arrangement of toggle clamps and its own clamping ring. These platforms have their central areas cut to meet different skeleton mould and sheet size requirements. At the base of the vacuum box is the air take-off point and a valve, which lead, ideally, to an air reservoir and vacuum pump.

The skeleton mould is mounted on the ram above the appropriate platform, which is firmly fixed to the vacuum box. The valve is closed on the vacuum box and the pump started (Figure 3.38A). A hot thermoplastic sheet blank is mounted on the platform and secured with the clamping frame and toggle clamps. The valve is opened to suck the air out of the vacuum box. When the plastics sheet Figures 3.38B and C has been pulled down sufficiently, the mould is lowered into it. When the mould is firmly in position (Figure 3.38D), the vacuum is slowly released, and as air enters the box the sheet tries to draw back. As it does so, it pulls taut over the mould, touching only at the corners.

This is a most useful extension of the press-forming technique. Once the basic structure and vacuum box have been made only top platforms and skeleton moulds are required to produce a wide variety of products not possible by other methods.

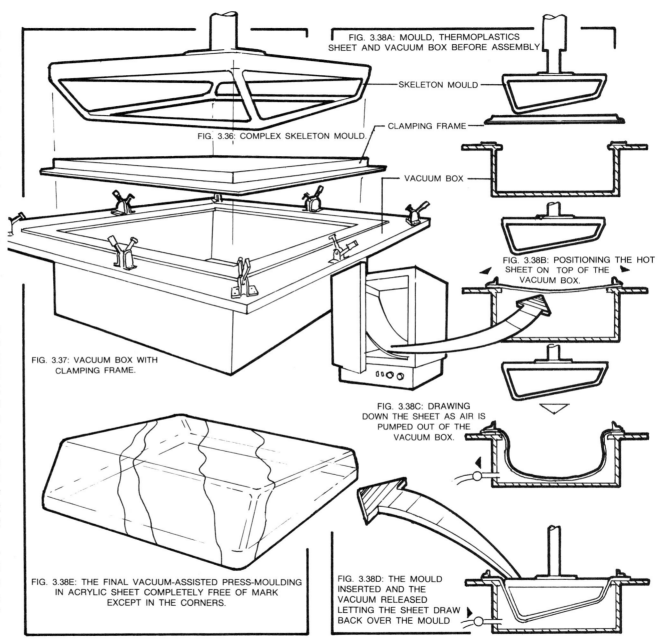

FIG. 3.36: COMPLEX SKELETON MOULD.

FIG. 3.37: VACUUM BOX WITH CLAMPING FRAME.

FIG. 3.38E: THE FINAL VACUUM-ASSISTED PRESS-MOULDING IN ACRYLIC SHEET COMPLETELY FREE OF MARK EXCEPT IN THE CORNERS.

FIG. 3.38A: MOULD, THERMOPLASTICS SHEET AND VACUUM BOX BEFORE ASSEMBLY

SKELETON MOULD

CLAMPING FRAME

VACUUM BOX

FIG. 3.38B: POSITIONING THE HOT SHEET ON TOP OF THE VACUUM BOX.

FIG. 3.38C: DRAWING DOWN THE SHEET AS AIR IS PUMPED OUT OF THE VACUUM BOX.

FIG. 3.38D: THE MOULD INSERTED AND THE VACUUM RELEASED LETTING THE SHEET DRAW BACK OVER THE MOULD

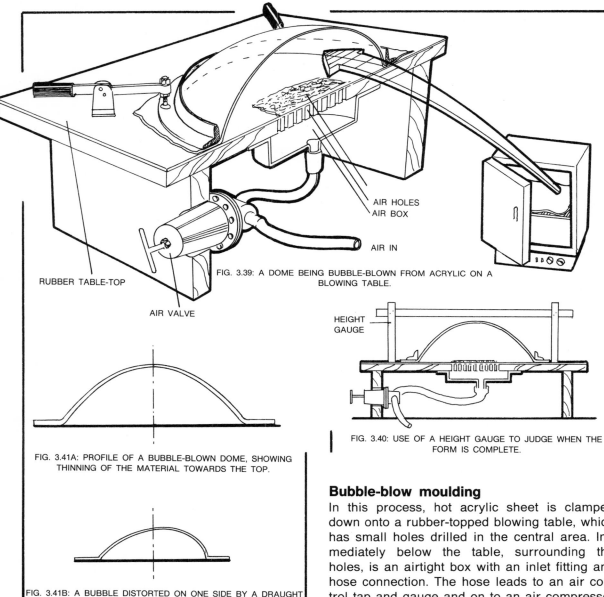

RUBBER TABLE-TOP

AIR VALVE

AIR HOLES
AIR BOX

AIR IN

FIG. 3.39: A DOME BEING BUBBLE-BLOWN FROM ACRYLIC ON A BLOWING TABLE.

FIG. 3.41A: PROFILE OF A BUBBLE-BLOWN DOME, SHOWING THINNING OF THE MATERIAL TOWARDS THE TOP.

FIG. 3.41B: A BUBBLE DISTORTED ON ONE SIDE BY A DRAUGHT DURING BLOWING.

HEIGHT GAUGE

FIG. 3.40: USE OF A HEIGHT GAUGE TO JUDGE WHEN THE FORM IS COMPLETE.

Bubble-blow moulding

In this process, hot acrylic sheet is clamped down onto a rubber-topped blowing table, which has small holes drilled in the central area. Immediately below the table, surrounding the holes, is an airtight box with an inlet fitting and hose connection. The hose leads to an air control tap and gauge and on to an air compressor (Figure 3.39). It is possible to use a car foot-pump and tyre valve if a compressor is not available. Once the material is securely clamped with a clamping ring and toggle clamps, air is blown into the box under the table, controlled by the valve. The air leaves the box through the holes in the table base, causing the hot thermoplastic sheet to rise and stretch. The bubble formed will continue to expand until the material has chilled and become hard again, or until the process is stopped by closing the valve and shutting down the air supply. Large bubbles of 3 mm-thick acrylic sheet need less air pressure than smaller bubbles made from the same thickness of material. A bubble will normally only become hemispherical — in other words, its height will not be greater than the plan radius. Cold draughts can chill a bubble on one side while it is forming, causing a distorted shape (Figure 3.41B). If holes in the base table are too large, they can allow cold jets of air to create chill marks on the bubble. A piece of gauze can be taped over any holes that cause chill marks.

Bubbles very rarely burst, but occasionally a weakness will appear in a sheet under pressure, causing a noticeable distortion. When this happens, cut off the air supply immediately.

A loud whistling noise indicates that the air is escaping; when this happens you should usually stop, readjust the clamping system, and reheat the blank before trying again. A fully blown hemispherical bubble will usually return to the flat if reheated. If a 3 mm-thick bubble is cut across its diameter it will be found to be 3 mm thick near the base clamping area, thinning progressively towards the top of the dome, which may be only 0.75–1 mm in thickness (see Figure 3.41A).

This process is most useful for producing protective exhibition dome covers, various hollow bowl shapes and some musical instrument hollow bodies (for example, lute, banjo or guitar).

For very small dome shapes it is better to press-form than to bubble-blow.

Note: The smaller the area of bubble, the greater the pressure needed to blow it. The larger the bubble diameter, the greater the volume of air needed, but with less pressure required, assuming a fixed thickness of acrylic.

Upside-down bubble-blowing table

Figure 3.42 shows a bubble being blown upside-down onto a piece of tube to make a 'stepped' dome. A rigid wooden frame structure is needed with a top plate and bottom surround that can be 'G' clamped firmly together. The top plate of wood contains a car tyre valve mounted in a steel disc with a rubber surface to make a seal.

The hot plastics sheet is then trapped between the top plate and a ring former (wooden shape) and clamped firmly together. A car foot pump is operated to blow a bubble. This technique is used to make flat bottomed domes and other shapes based on hemispheres.

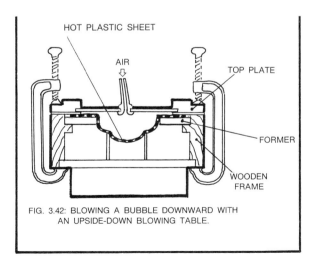

FIG. 3.42: BLOWING A BUBBLE DOWNWARD WITH AN UPSIDE-DOWN BLOWING TABLE.

PROJECTS AND ACTIVITIES

(1) You are to make a triangular box using a strip bender. Design a jig that will enable you to do this, and sketch its shape. Why is it necessary to plan the order in which strip bends are carried out?

(2) Using the strip bender for the moulding method, design a product to support 10 paperback books or 10 cassettes for a tape recorder.

(3) Design a light fitting suitable for use in a 3–9-year-old child's bedroom, using strips of clear PVC sheet formed by the strip-bending process.

(4) Take a selection of pieces of different types of thermoplastics sheet, all of the same thickness. Describe the 'feel' of each material. Apply heat and measure the time it takes for each one to soften to moulding temperature.

(5) Design an acrylic cover for a porch light, or a cover for lights to illuminate a footpath at your home or school. Is this the most suitable material? What are the environmental problems you should consider? Suggest

A sheet bubble blowing table showing clamping ring and toggle clamp holding system. The bubble being blown is 140 mm diameter and in acrylic sheet.

shapes suitable for making the cover using the press-forming technique. Describe the moulding technique and draw the mould.

(6) You have just bought a new camera or calculator. Design a press-formed carrying case in which to keep it. Suggest a suitable material, and illustrate your design and the mould(s).

(7) Your school needs a light box in the CDT area for examining slides and preparing tracings from small (A3-size) drawings. Research the electrical and safety requirements for this equipment and prepare a design, describing moulding and assembly techniques.

(8) It will soon be your mother's birthday and you are short of pocket money to buy her a present. Design and make her either a small ring box or a bracelet, using the principle of 'plastic memory'. The lid of the box should be decorated and the sides and bottom can be based on tube or sheet as required.

(9) The hotter a piece of thermoplastic material becomes, the lower its tensile strength becomes. Devise an experiment to test this, and then draw a table or graph showing tensile strength versus temperature for one material. If others in the class choose different materials, make a comparison between materials that could be useful in designing products that have to work at above normal room temperatures.

(10) The thin plastics material used for coffee cups that come from drinks machines often allow one's fingers to be burned. Design a holder for these cups.

(11) Plan and carry out an experiment to find out how poor samples of different plastics are at conducting heat. Does this have any effect on their processing?

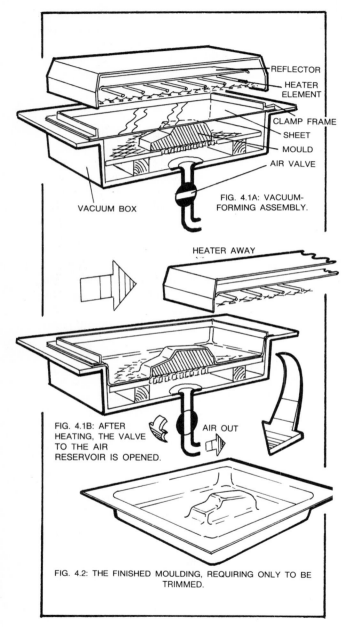

FIG. 4.1A: VACUUM-FORMING ASSEMBLY.

REFLECTOR
HEATER ELEMENT
CLAMP FRAME
SHEET
MOULD
AIR VALVE
VACUUM BOX

HEATER AWAY

FIG. 4.1B: AFTER HEATING, THE VALVE TO THE AIR RESERVOIR IS OPENED.

AIR OUT

FIG. 4.2: THE FINISHED MOULDING, REQUIRING ONLY TO BE TRIMMED.

Vacuum forming is another thermoplastics sheet-moulding process, but it differs from the processes described in Chapter 3 because it requires a special type of machine. The plastics material has to be clamped in position over an open box and heated. When hot, the soft material is easily moulded by pumping the air from the box. The hot sheet then takes up the shape of the box interior (to form a product like a bath), or it takes the shape of a mould placed in the box.

The process is a good one for making small mouldings from thin materials. For example, if your local model racing-car club asked you to make 500 model car bodies, you could make them by vacuum forming, using thin sheet thermoplastics. The model-makers could fit their own chassis and power units to the body shells.

THE PROCESS

A simple vacuum-forming machine requires only a vacuum pump (a vacuum cleaner makes a good one), an airtight box, a strong mould, a clamping frame, and a heater system (Figure 4.1A). A valve and air reservoir complete the system.

The thermoplastic sheet is clamped and heated from above. When it is hot, the heater is removed (Figure 4.1B). The vacuum pump is run at the same time as the heater to create a vacuum in the reservoir, but the valve is kept closed. Then the valve is opened and air rushes from the vacuum box to the reservoir. The pump continues to run to remove that air from the reservoir.

Atmospheric pressure above the plastic sheet pushes the hot sheet down. The mould is in the box, just below the sheet, so the plastics material drapes over it, creating a moulding. The material, being thin, chills quickly and the moulding may be removed for trimming and other work (Figure 4.2). This process is known as *drape* vacuum-forming.

Making a mould for vacuum forming

At the moment of forming, the mould has a vacuum below it. Atmospheric air pressure then weighs very heavily on the mould; a weak mould can be completely crushed. The mould must be made rigidly from strong materials. Hollow objects and thermoplastic materials are not suitable for moulds. Small moulds are best made from blocks of wood, or cast in plaster.

When deciding on how to design a mould, first decide whether it should be a positive or negative form. Make this decision from the following points:

(1) which side of the finished moulding is most important (will be seen);
(2) which mould form will allow the material to stretch easily and give the greatest thickness where it is most needed; and
(3) which mould form is the simplest to make.

Vacuum forming moulds and moulding used for producing a model of electronic weighing scales for blind people.

The mould surface must, of course, be very good. Even with a good surface the soft plastic material can be marked by woodgrain pattern, by fine particles of dust and by tiny flaws in the mould, such as joint lines. An otherwise good moulding can be spoilt by these marks, which cannot easily be removed.

The Foremech 450 vacuum forming machine with one pair of reducing plates in position. The three temperature controls for heating the zoned areas can be seen in the moveable heater head.

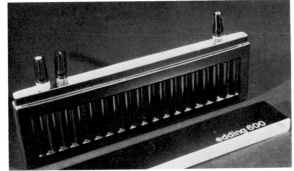

Two identical vacuum formed mouldings have been used to produce this container for Edding felt pens. The mouldings were trimmed and cemented together before being sprayed with an etch primer and top coat. The material is high impact polystyrene.

Negative or hollow areas in the mould surface must be vented to allow the air to escape. Mould design points and venting are detailed on pages 43 and 44.

THE AIR-SLIP TECHNIQUE

The moving-table machine

A special machine (Figure 4.3) is required for this process, which is a more modern version of the simple technique described above. The advantages that it offers are:

(1) deeper, more complex mouldings may be made;
(2) a better distribution of material is possible;
(3) only a small air reservoir is necessary;
(4) it is quicker, because only a small amount of air needs removing.

The moving-table vacuum-forming machine has a mould table that can be moved up and down, either by compressed air or by a mechanical rack and pinion system. When the table is up, its surface is about 10 mm from the underside of the plastic sheet. The volume of air in this space is very small, and since the table-top seals against an inward-facing flange at the top of the vacuum chamber, it is only this small quantity of air that needs evacuating to the reservoir, which is both quick and efficient.

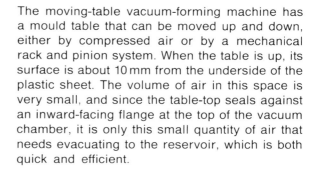

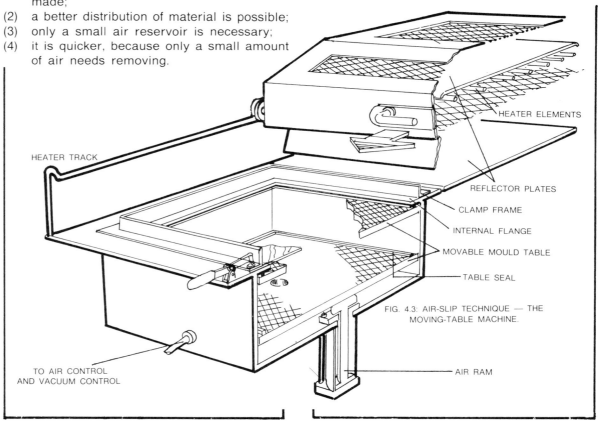

HEATER ELEMENTS

HEATER TRACK

REFLECTOR PLATES

CLAMP FRAME

INTERNAL FLANGE

MOVABLE MOULD TABLE

TABLE SEAL

FIG. 4.3: AIR-SLIP TECHNIQUE — THE MOVING-TABLE MACHINE.

AIR RAM

TO AIR CONTROL AND VACUUM CONTROL

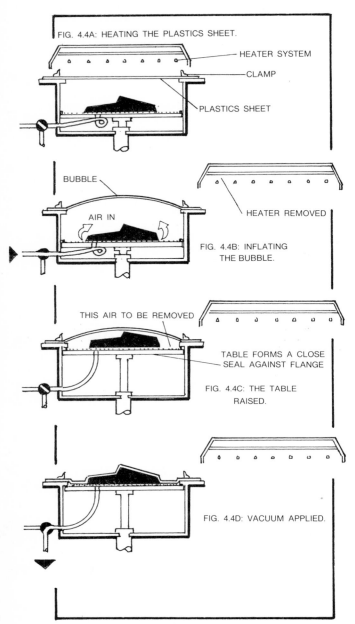

FIG. 4.4A: HEATING THE PLASTICS SHEET.
HEATER SYSTEM
CLAMP
PLASTICS SHEET

BUBBLE
HEATER REMOVED
AIR IN
FIG. 4.4B: INFLATING THE BUBBLE.

THIS AIR TO BE REMOVED
TABLE FORMS A CLOSE SEAL AGAINST FLANGE
FIG. 4.4C: THE TABLE RAISED.

FIG. 4.4D: VACUUM APPLIED.

When the table is down, it may be 150 mm below the plastic sheet, or as much as 300 mm on some machines. The depth to which the table may be lowered determines the maximum depth of mould that may be used. It is most unusual for a vacuum forming to be made more than 100 mm deep from machines with small surface areas (that is, less than 500 mm × 400 mm).

Bubble preforming

Position the heater hood over the plastics sheet (Figure 4.4A), after placing the mould on the moveable table. Warm the sheet through slowly with the heater. When it is hot enough, the sheet is soft and moves easily if touched. Move the heater away and blow air into the vacuum box, causing the hot sheet to swell up into a shallow bubble (Figure 4.4B).

Raise the moving table to bring the mould up into the bubble (Figure 4.4C). Once the mould is in the bubble, start the vacuum pump and pump out the air from the space around the mould. The plastics sheet then takes up the shape of the mould (Figure 4.4D). After a minute or two the moulding is sufficiently rigid to be removed and trimmed.

Various types of packaging can be produced by vacuum forming.

Blister packaging

Thin clear PVC sheet is vacuum formed over a solid mould that represents the shape of the product to be displayed. The product is then placed on a piece of printed card coated with a heat softening adhesive. The vacuum-formed clear blister is placed over the product and the card and a hot iron used to heat seal the moulding to the card to hold the product in position for display.

Skin packaging

This process uses vacuum forming to draw a very thin plastic skin down over the product onto perforated or porous card. The plastic skin welds to a coating on the card surface trapping the product. This is often used to hold several items together as part of a display.

Flocked packaging

Some plastics sheet materials (most notably polystyrene sheet) are available coated with Nylon flock (fine hairs giving a velvet type of surface finish). These are available in a wide range of colours and textures. A mould is placed on the vacuum forming machine with the flocked material placed with the flock (hairy surface) uppermost. It is then vacuum formed in the usual manner and the finished moulding can be used to display precious objects, such as watches.

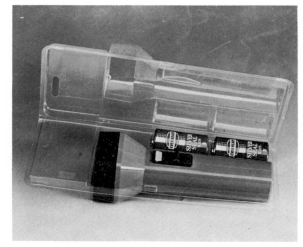

Display 'blister-pack' vacuum formed from PVC (thin rigid sheet).

Mould design for vacuum forming

The diagrams on this and the following page show how mould design can affect the distribution of thickness of material in the finished vacuum-formed moulding. They also show simple methods for venting moulds in concave corners, and the need for trimming allowances.

In Figure 4.5, the hot plastics material will touch the top surface of the cold mould and chill, but the area over the main cavity will continue to draw down, thinning as it goes. Because this area of sheet (distance x) is large in relation to the depth, it will need to stretch to only twice its original length and thin to approximately half its original thickness. In Figure 4.6, however, the surface area of the sheet (distance y) is smaller in relation to the cavity depth. In this case the sheet will have to stretch to four times its original length as it moulds round the cavity, and will finish with a quarter of its original thickness. This means that to make this moulding you must either start with thicker sheet, making the moulding more expensive, or be prepared to accept a very thin corner at the bottom of the depression. Overleaf the same moulding is shown being made, first from a negative mould (Figure 4.10) and then from a positive mould (Figure 4.11). The mould at the top of the page will produce a more even material thickness throughout.

Venting is essential in the concave corners of the moulds to allow air to escape. Venting or air holes should be very fine — 1.5 mm or less in diameter at the mould surface (Figure 4.7); otherwise they will 'print' on the hot sheet. Where moulds are built up in layers, large holes can be drilled under the positive forms and thin pieces of postcard used to separate the layers; the evacuated air is then free to escape quickly between the laminations (Figure 4.8 and 4.9).

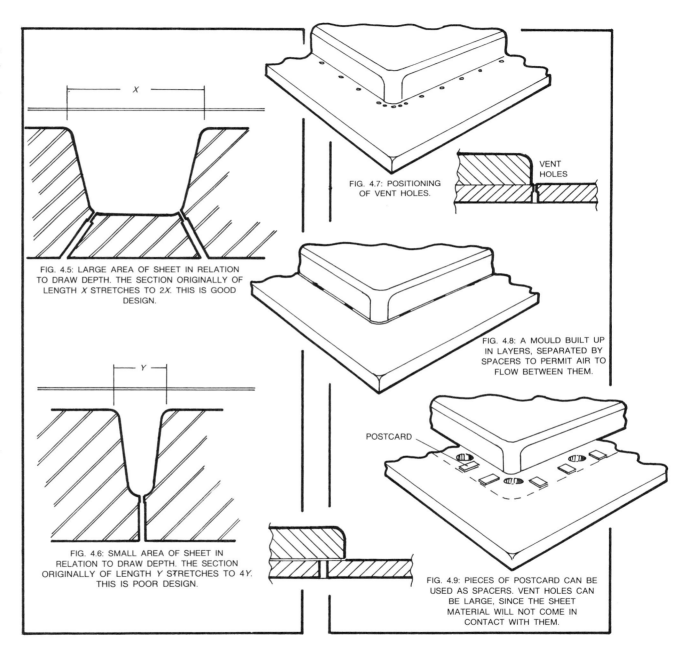

FIG. 4.5: LARGE AREA OF SHEET IN RELATION TO DRAW DEPTH. THE SECTION ORIGINALLY OF LENGTH X STRETCHES TO $2X$. THIS IS GOOD DESIGN.

FIG. 4.6: SMALL AREA OF SHEET IN RELATION TO DRAW DEPTH. THE SECTION ORIGINALLY OF LENGTH Y STRETCHES TO $4Y$. THIS IS POOR DESIGN.

FIG. 4.7: POSITIONING OF VENT HOLES.

VENT HOLES

FIG. 4.8: A MOULD BUILT UP IN LAYERS, SEPARATED BY SPACERS TO PERMIT AIR TO FLOW BETWEEN THEM.

POSTCARD

FIG. 4.9: PIECES OF POSTCARD CAN BE USED AS SPACERS. VENT HOLES CAN BE LARGE, SINCE THE SHEET MATERIAL WILL NOT COME IN CONTACT WITH THEM.

43

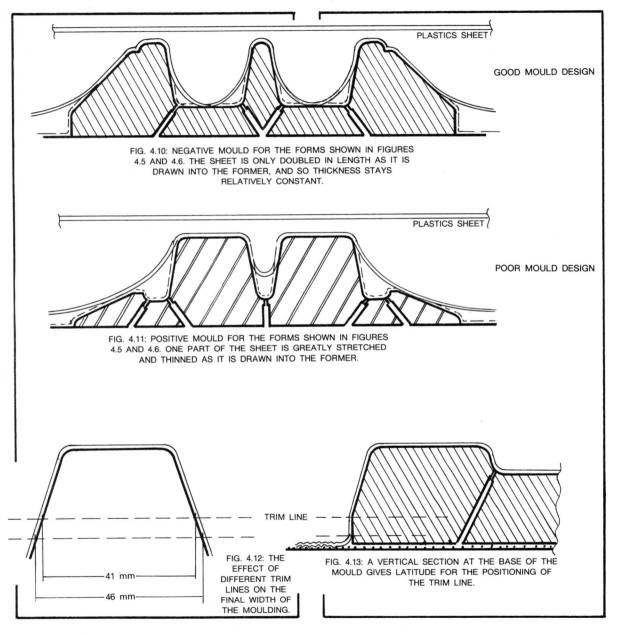

FIG. 4.10: NEGATIVE MOULD FOR THE FORMS SHOWN IN FIGURES 4.5 AND 4.6. THE SHEET IS ONLY DOUBLED IN LENGTH AS IT IS DRAWN INTO THE FORMER, AND SO THICKNESS STAYS RELATIVELY CONSTANT.

PLASTICS SHEET

GOOD MOULD DESIGN

PLASTICS SHEET

POOR MOULD DESIGN

FIG. 4.11: POSITIVE MOULD FOR THE FORMS SHOWN IN FIGURES 4.5 AND 4.6. ONE PART OF THE SHEET IS GREATLY STRETCHED AND THINNED AS IT IS DRAWN INTO THE FORMER.

TRIM LINE

41 mm
46 mm

FIG. 4.12: THE EFFECT OF DIFFERENT TRIM LINES ON THE FINAL WIDTH OF THE MOULDING.

FIG. 4.13: A VERTICAL SECTION AT THE BASE OF THE MOULD GIVES LATITUDE FOR THE POSITIONING OF THE TRIM LINE.

The choice of whether a mould is made as a positive or negative form depends on:
(1) the complexity of the shape required;
(2) which surface of the moulding is most likely to be seen (because of mould marks);
(3) the depth of draw in relation to the surface area;
(4) mould-making materials and methods available.

All moulds must have radiused corners, otherwise the plastics material will be punctured. The sides must be tapered — ideally at 3° on each side. The surfaces should be gently curved where possible. Square positive corners should not be close to one another, otherwise the plastics material will 'web' (Figure 4.14).

Mouldings have to be trimmed — that is, the sheet around the moulding has to be removed. It is wise to make the base of the mould vertical at its lower edges to allow for trimming (Figure 4.13). Tapered mouldings become narrower when they are trimmed at different depths — a simple point often overlooked (Figure 4.12).

When mouldings are trimmed with a saw or a knife, great care must be taken to avoid splitting or cracking. When trimming with a knife, it is sometimes helpful to replace the moulding over the mould to provide a firm support.

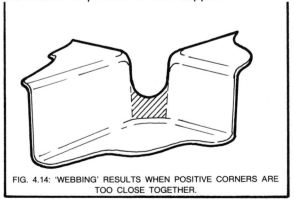

FIG. 4.14: 'WEBBING' RESULTS WHEN POSITIVE CORNERS ARE TOO CLOSE TOGETHER.

PROJECTS AND ACTIVITIES

(1) Take an empty coffee-machine plastic cup, measure its thickness and calculate its surface area from its height and diameter. Heat it until it softens and 'demoulds' (return to being a flat sheet), and allow it to cool. When it is cold, measure the thickness and area of the material. Compare this with the original thickness and surface area. How much did the material stretch and thin to make the cup?

(2) Design a model boat or vehicle whose motive power comes from the wind. Select and use any plastics materials and processes you think appropriate.

(3) Using vacuum forming as the moulding process, prepare product and mould design ideas for one of the following:-
1. A model racing car, electrically powered for track racing.
2. A pencil, pen and pin tray drawer liner.
3. A needlework box drawer liner.
Measure and closely examine the components or products to be fitted into your moulding.

(4) Some plastics materials can withstand impact better than others. Devise a test to find out the impact resistance of small samples of different plastics at: room temperature; 0°C; 40°C.

(5) Design a rig to test how resistant common plastics are to abrasion (wear from rubbing).

(6) Using the principle of thin-sheet vacuum-forming (and possibly strip-bending as a follow up process) design and make a holder for your cassette tapes.

(7) Design and make a mechanical or electronic calendar that can be attractively housed in a vacuum formed case made from Foamex PVC sheet.

(8) Take a 1mm thick sheet of high impact polystyrene of suitable area to fit your vacuum-forming machine and a mould of hemispherical shape 100mm in diameter. (A mould may be made by making a plaster cast from the inside of an old PVC or rubber ball.) Mark out the plastics blank with 10mm squares and vacuum form the sheet over the hemispherical mould. Then examine and measure the amount of distortion that has taken place over the surface by noting the change of size of the squares. Finally carefully cut the squares out and using a micrometer measure the thickness of material. Make a table showing amount of distortion in relation to change of thickness.

(9) Most modern houses have low ceilings making overhead central light fittings a problem in hallways. Prepare a range of design proposals that exploit sheet vacuum forming of thin PVC and the new low heat output light fittings. (Look at the Thorn 2D lamp, for example, which can be mounted against the ceiling.) If you decide to use conventional light bulbs as an alternative allow plenty of space for air to circulate between the bulb and the moulding.

(10) Using vacuum forming prepare a package design suitable for the protection and display of a set of Christmas tree lights. Consider using screen printing on a clear PVC vacuum formed lid. Mark your design out on a formed lid, reheat the material and use plastic memory to take it back flat under the vacuum forming machine — this will give you the distorted shape for the screen layout. Then screen print each panel flat before vacuum forming. When vacuum formed the pattern should lose its distortion and come out correctly shaped. (Be careful to locate the mould correctly in relation to the sheet.)

(11) Do you ever lose the soap, sponge and other accessories in the water when having a bath? Design a device to house these items that can be made by vacuum forming and that will fit on to the side of the bath.

(12) Using your knowledge of simple mechanisms and vacuum forming of sheet materials design and make a mechanical toy which explodes, falls apart, or ejects its occupants when it crashes.

(13) Visit a local historical country mansion open to the public and ask the owner/manager if you could design a product that might be sold as a souvenir. A typical product might be a relief wall plaque showing the outline of the building, some words and numbers stating its name, motto, and date. Design the unit for vacuum forming and prepare full costings taking into account material and mould cost, time and labour cost, packaging and transport, profit and retail selling price.

(14) Have a discussion with your school geography teacher and find out if he or she could usefully use a relief vacuum-formed contour map of your area. Decide on the landscape features and details. Then prepare a design that would permit the map to be wall mounted and if possible fit a low voltage lighting system to enable pupils to locate quickly places of importance. The map could be made in sections if it is larger than your machine and additional mouldings could be prepared to house rock samples, show sections and rock strata between specific points.

5 JOINING PLASTICS

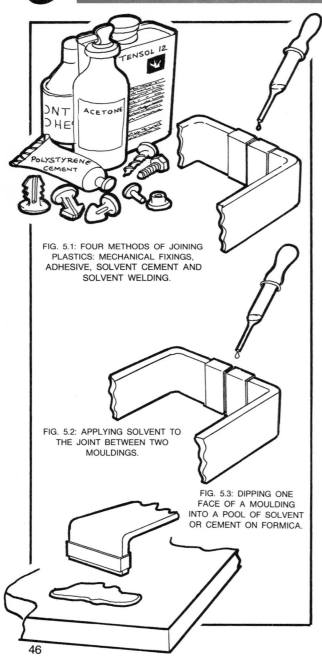

FIG. 5.1: FOUR METHODS OF JOINING PLASTICS: MECHANICAL FIXINGS, ADHESIVE, SOLVENT CEMENT AND SOLVENT WELDING.

FIG. 5.2: APPLYING SOLVENT TO THE JOINT BETWEEN TWO MOULDINGS.

FIG. 5.3: DIPPING ONE FACE OF A MOULDING INTO A POOL OF SOLVENT OR CEMENT ON FORMICA.

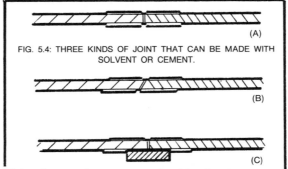

(A)

(B)

(C)

FIG. 5.4: THREE KINDS OF JOINT THAT CAN BE MADE WITH SOLVENT OR CEMENT.

This chapter is concerned with simple ways of joining plastics materials and mouldings. The methods shown are suitable for use in schools and do not require the extremely expensive machines used by industry. Four main methods are outlined, and each has many variations. At least one of the methods will suit all the common plastics in school use. Figure 5.1 shows the preferred methods.

Mechanical fixings

Generally these are most suitable for rigid plastics mouldings and are often used for components that have to be dismantled at a later date for the replacement of parts (such as batteries) or for maintenance. In some cases parts can be made to screw together or clip into one another. Where these methods are not suitable, self-tapping screws and bolts can be used, provided they are removed only occasionally. Plastics need coarse threads if they are to wear well; small threaded brass inserts should be fitted if constant assembly and dismantling are to be expected. Rivets are often used as a permanent method of mechanically fixing parts.

Wherever possible, either make a deliberate visual feature of a mechanical fixing or keep it hidden on the back or underside of the product. Avoid nuts and bolts showing on important faces.

Adhesives

Warning: Many adhesives are highly flammable and give off fumes. Do not use them near a naked flame.

There are many safe commercial adhesives on the market that are suitable for joining different types of plastics. Some of these adhesives have a permanent 'rubbery' quality, which makes them particularly suitable for joining flexible plastics to each other and to other materials. There are 'hot melt' adhesives which require heat to melt a plastics pellet or film, and there are various double-sided coated films and foams that only require the protective skin to be removed. Some adhesives rely on the evaporation of a solvent, while others are prepared by mixing two liquids or pastes so that they are 'cured' by chemical reaction.

Always match the adhesive to the materials you wish to join together. Some adhesives are dangerous and not suitable for use in schools, although some are quite common at home. Epoxy resin-based adhesives are poisonous. Other adhesives, if mishandled, will quickly stick fingers together (for example, cyanoacrylates). Most adhesives manufacturers supply charts

Plastic fixings.

showing which of their preparations are most suitable for joining a wide variety of different types of plastics and non-plastics materials, and advice on the best method to use to make joints.

SOLVENTS AND CEMENTS

Many thermoplastics materials can be joined to themselves using solvents or solvent cements. Some solvent cements work on several plastics, while others work on only one type. This makes it important to choose the most suitable cement, especially when joining two different plastics.

A solvent is usually a liquid that softens and/or dissolves a plastics material. The solid plastics material then becomes temporarily liquid, owing to the solvent action. Two pieces of thermoplastic material may be joined by dissolving their edges in a solvent, lightly pressing them together and leaving the assembly while the solvent evaporates. Then the softened joint solidifies and the two materials become joined.

A solvent cement is made up of plastic material dissolved in a solvent to form a paste. Paste cements work in much the same way as the pure solvent but the paste has the advantage of filling any fine gaps in the joint, and is simpler to apply. However, if too much cement is used it will shrink back unevenly.

Warning: If solvent cements are to be made in school they should be prepared under strict supervision in laboratory conditions. Many solvents are highly flammable and must not be used near naked flames. The fumes can be harmful when breathed in, in any concentration, even for a short period of time. Therefore all solvent and solvent cement work should be done in a fume cupboard or well ventilated area. Use the minimum of solvent or solvent adhesive at any one

time. With manufacturers' proprietary cements, follow all instructions exactly.

Solvents will evaporate first from the surface of the joint and more slowly from the centre of the joint. Thin material is therefore more suitable for this process than thick sheet.

Figure 5.2 shows an eye-dropper about to apply solvent to the join line between two masked mouldings. When the solvent has been in the joint for a few moments and the plastics have softened, apply slight pressure to make sure that the materials come together firmly. *Do not leave the mouldings under pressure but lightly hold them together with adhesive tape.* Mouldings held under pressure develop internal forces, which the chemicals in the cement attack. These lines of force show in the mouldings later as silvery internal cracks.

Figure 5.3 shows a small pool of solvent or cement on a Formica board with a masked plastic moulding about to be dipped into it. When the edge has been lightly coated it may be transferred to an assembly board and lightly pressed into contact with the second moulding. Again, the mouldings should be left without further pressure being applied.

The different types of joint that you can make by this technique are sketched in section form in Figures 5.4A–C.

Immediately the joint has been made, the remaining pool of solvent cement should be wiped up and the rag disposed of *outside* the building *in a dustbin*.

Warning: Solvent-impregnated rag is a fire risk and if left lying around in a workshop may eventually ignite through spontaneous combustion (that is by itself).

Figures 5.5A–D and 5.6 show the preparation, making and supporting of a T-butt joint.

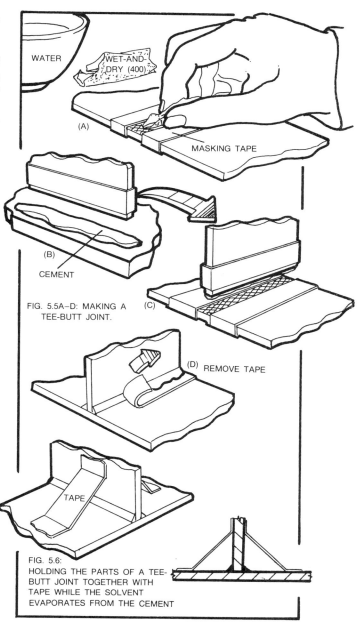

FIG. 5.5A–D: MAKING A TEE-BUTT JOINT.

FIG. 5.6: HOLDING THE PARTS OF A TEE-BUTT JOINT TOGETHER WITH TAPE WHILE THE SOLVENT EVAPORATES FROM THE CEMENT

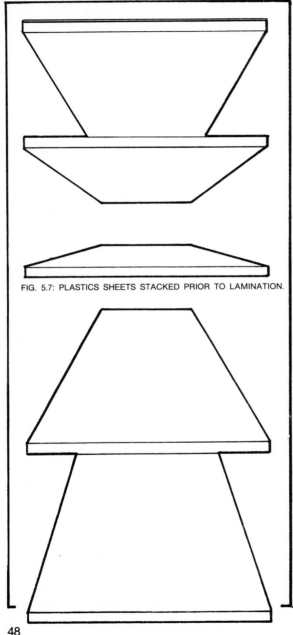

FIG. 5.7: PLASTICS SHEETS STACKED PRIOR TO LAMINATION.

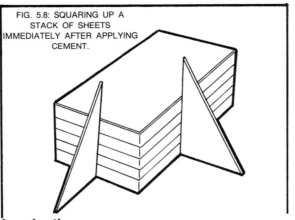

FIG. 5.8: SQUARING UP A STACK OF SHEETS IMMEDIATELY AFTER APPLYING CEMENT.

Lamination

Laminating is joining sheets together, face to face, as in Figure 5.8. A 'two-part' cement made by ICI, called Tensol 70, is the best adhesive.

Warning: This must be mixed by your teacher. Wear disposable polythene gloves to handle the mix, and keep the mix and the laminated sheets in a fume cupboard until cured.

The two parts are (a) methyl methacrylate monomer, containing '*mers*' — the building blocks for this plastics material — and (b) benzoyl peroxide, which causes the individual mers to link to form the polymer. Bringing the two parts together in the correct proportions makes polymethyl methylacrylate (PMMA), the cement being the same as the plastics being used.

A solvent cement is not suitable for making sheet laminations, as the solvent cannot evaporate from the centre of the joined surfaces. Tensol 70 is more suitable because it cures by a chemical reaction.

To laminate a series of acrylic sheets of different colours, the method is as follows. Mark and cut out the pieces of sheet with the protective paper in place. Remove the paper from both sides of each sheet and lightly file the edges to remove any burrs. Wash and dry the pieces thoroughly.

Assemble a stack so that the lamination arrangement can be checked. Apply masking tape over the entire top surface of the top piece and bottom surface of the bottom piece. The remaining surfaces may now all be lightly rubbed down with damp wet-and-dry abrasive paper (400).

When all the surfaces have had the gloss removed, they should be washed again to remove every trace of dust, and then dried. Assemble the sheets in a line in the stacking order, pairing the faces to be cemented.

Ask your teacher to measure out the Tensol 70 cement components, and to mix them. Allow the mix to stand for about a minute to allow any bubbles to come to the surface. Remove as many of these bubbles as possible by drawing them to one side.

Pour a pool of cement into the centre of the bottom sheet (Figure 5.9A). Lay the next sheet on top and press the two firmly together until the cement stops oozing from the joint. Apply a pool of the cement to the centre of the second sheet, apply the next layer, and so on. Figure 5.9A shows the correct method to apply cement; Figure 5.9B shows the incorrect method, which will trap air.

Finally, square up the layers as shown in Figure 5.8.

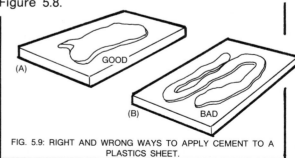

FIG. 5.9: RIGHT AND WRONG WAYS TO APPLY CEMENT TO A PLASTICS SHEET.

WELDING

Some thermoplastics cannot be dissolved easily in 'safe' solvents — for example, polyethylene and polypropylene. Some joints can be cemented but are better heat-welded. To join these materials it is usual to use heat and pressure.

There are several methods of applying heat, depending on the material and its thickness:

(1) a rotary welding tool, which is similar to an electric soldering iron (Figure 5.11);
(2) a plate 'mirror' welder, or the sole plate of a domestic iron (Figure 5.14);
(3) a hot-air gun (Figures 5.12, 5.13, 5.15);
(4) an oven for sheet lamination (Figure 5.16);
(5) a bag welder for film (Figure 5.17);
(6) friction welding (Figure 5.18);
(7) high-frequency welding;
(8) ultrasonic welding.

Methods 1 to 6 are available for use in many schools and colleges. As processes they are straightforward and important for the making of many plastics articles. Methods 7 and 8 are important industrial processes.

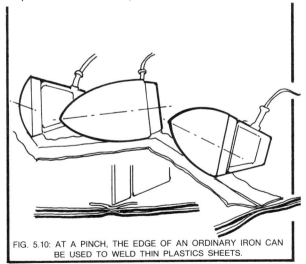

FIG. 5.10: AT A PINCH, THE EDGE OF AN ORDINARY IRON CAN BE USED TO WELD THIN PLASTICS SHEETS.

Using a rotary welder

The rotary welding tool is a device similar to an electric soldering iron, except that, instead of a copper tip, it has a 15 mm-diameter chromium-plated wheel, which is free to rotate. The shaft supporting the wheel houses an electrical element. Heat from the element is conducted to the wheel, which rises in temperature to the melting range of most thermoplastics (Figure 5.11).

Place a heat-resistant clear plastic cooking film sheet over a masked-out double thickness of polyethylene film. Put the sheets on a heat-resistant surface. When the welding tool is hot, use it to slowly trace round the marked-out shape on the polyethylene. The cooking film prevents the hot tool from completely melting or cutting through the plastic film. The speed and pressure that you work at will affect the quality of the welded joint. If the tool passes too quickly, there will be insufficient time for the heat fully to melt the two layers of film together. If it moves too slowly, then the material will melt, thin, and lose its strength. A few minutes' practice on some offcuts of the same material is all that is necessary to determine the correct speed.

The advantage of this tool is that it can be made to follow tight curves. Its disadvantage is that it can be used only on thin film. A variety of roller wheel profiles are available for different purposes. The rotary tool can be used for making up all sorts of plastic wallets and bags. It is also possible to make kites and thin inflatable structures.

A hot-air welder with a wheel attachment can be used in the same way (Figure 5.12).

If the correct equipment is not available, a soldering iron or domestic iron can be used (see Figure 5.10).

Warning: Never allow the materials to burn or give off smoke!

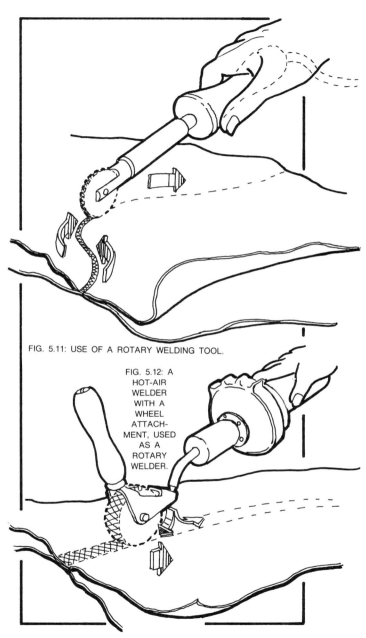

FIG. 5.11: USE OF A ROTARY WELDING TOOL.

FIG. 5.12: A HOT-AIR WELDER WITH A WHEEL ATTACHMENT, USED AS A ROTARY WELDER.

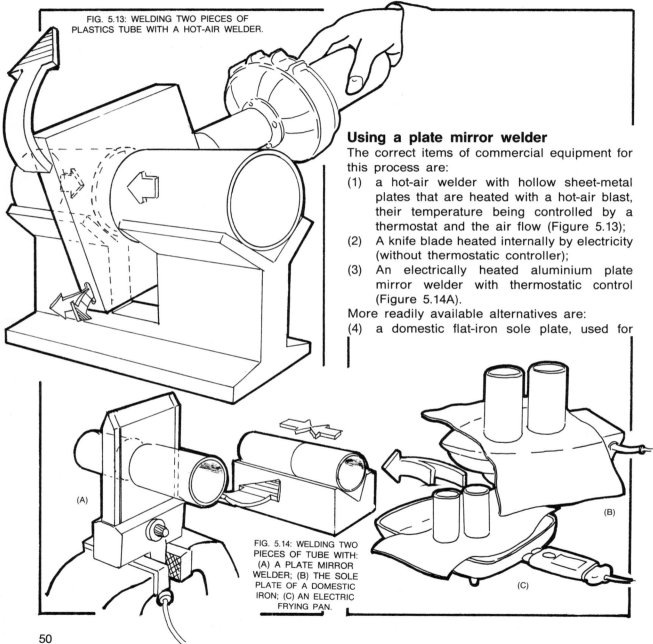

FIG. 5.13: WELDING TWO PIECES OF PLASTICS TUBE WITH A HOT-AIR WELDER.

(A)

FIG. 5.14: WELDING TWO PIECES OF TUBE WITH: (A) A PLATE MIRROR WELDER; (B) THE SOLE PLATE OF A DOMESTIC IRON; (C) AN ELECTRIC FRYING PAN.

(B)

(C)

Using a plate mirror welder

The correct items of commercial equipment for this process are:

(1) a hot-air welder with hollow sheet-metal plates that are heated with a hot-air blast, their temperature being controlled by a thermostat and the air flow (Figure 5.13);

(2) A knife blade heated internally by electricity (without thermostatic controller);

(3) An electrically heated aluminium plate mirror welder with thermostatic control (Figure 5.14A).

More readily available alternatives are:

(4) a domestic flat-iron sole plate, used for ironing clothes — not a steam iron (Figure 5.14B);

(5) an electric frying pan with a cast-in element in its base (Figure 5.14C).

All of the above items need either a PTFE (Teflon) type of non-stick surface on the welder face, or non-stick sheet material between the hot plate and the moulding during the heating operation. This is to prevent material melting, sticking to the surface and burning.

Supporting frames or jigs are necessary to hold the mouldings in alignment after heating, when they are being brought together to obtain a good joint of maximum strength. For tube, for example, the jig may be a vee-jig — see Figures 5.13 and 5.14.

Set the welder at a temperature well above the melting temperature of the plastic moulding. This is to make a small area melt and become liquid quickly. If the welder temperature is set only to the melting temperature of the material, the longer time necessary for heating will cause the moulding to soften for several millimetres from the weld point. When pressed together the softened material spreads out to form a large flange. This can be unsightly, but does give a very strong joint.

Heat two pieces of material together, watching them closely. As soon as the material has become molten, bring the mouldings together in the jig and hold them firmly until cold. Speed is essential, for the material will remain molten for only a few moments once it has left the heater. This weld will not be as strong as the material itself.

The advantage of this method is that it is quick. But if it is mishandled, it produces a dirty weld due to burnt material. Other disadvantages are that it is suitable only for welds in one plane and that the weld size is limited by the area of the welder hot-plate.

Hot-air welding

Plastics can be joined with a hot-air welder equipped with a nozzle for a *welding rod*. The welding rod is a filler for the joint, consisting of the same material as the sheet to be welded. It must seat well down in the joint. To help the welding rod seat properly and to heat thick mouldings thoroughly it is essential to cut back the joint area to allow rapid heat penetration. A coarse file is the best method of producing a *chamfer* on the joint line. Very thick material needs a double chamfer, creating a knife-edge section. Figure 5.15A shows different chamfer methods.

Sheets are clamped into their correct positions. The clamps must not interfere with the welding line. When operating the welder, set it at a temperature well above the melting temperature of the material. The quality of weld is then controlled by the speed at which the welder is moved across the joint. If any material burns, the welder is moving too slowly, while if the welding rod or the joint does not soften properly, the welder is being moved too fast. It is wise to practise with offcuts before attempting the finished product weld.

You need both hands to make the weld: one to hold the machine, the other to feed in the welding rod. Start the machine and set the temperature on the rotary switch in the handle. Put the end of the welding rod into the nozzle guide so that 12 mm of material sticks out from the end of the nozzle. When the hot airflow softens the welding rod, the machine is at working temperature and ready for use. The nozzle base is designed to preheat the material weld line before the welding rod is pressed into position. *The machine is drawn backwards* and lays hot welding rod on to the heated weld line of the moulding (Figure 5.15B). At first this feels an unnatural movement, but it is the only way to

create a perfect weld. For thin materials or mouldings only one pass of the welder may be required, while for double-chamfered thick materials as many as three passes per side (putting down three layers of welding rod in each) may be necessary.

Figure 5.15C shows the nozzle shape, airflow and line of movement in detail. The advantages of this method are:

(1) thick materials may be effectively welded together;
(2) the weld line need not be in only one plane but may follow a curved line;
(3) tee-joints may be made on flat sheet without heating large areas, which can cause distortion.

The disadvantages are:
(1) the process requires practice and careful co-ordination;
(2) the joint may look untidy, and needs to be out of sight where possible.

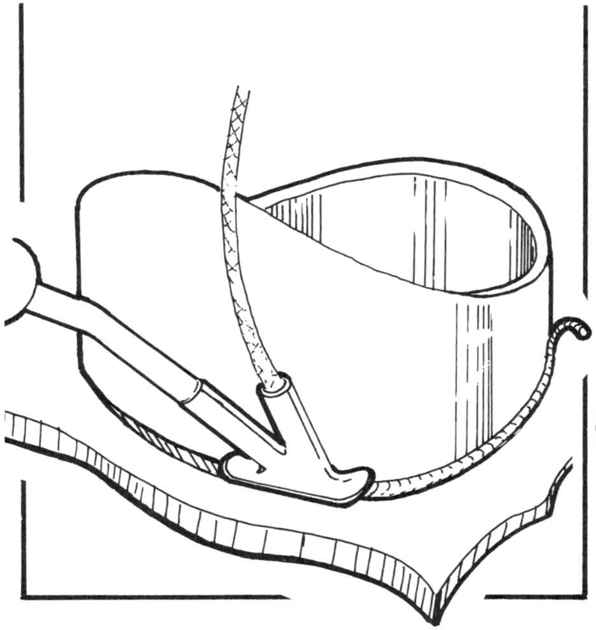

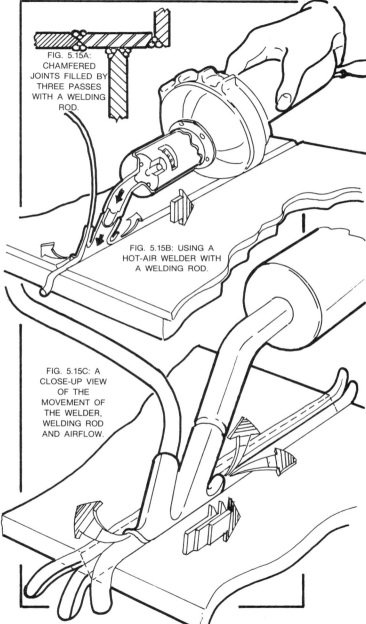

FIG. 5.15A: CHAMFERED JOINTS FILLED BY THREE PASSES WITH A WELDING ROD.

FIG. 5.15B: USING A HOT-AIR WELDER WITH A WELDING ROD.

FIG. 5.15C: A CLOSE-UP VIEW OF THE MOVEMENT OF THE WELDER, WELDING ROD AND AIRFLOW.

51

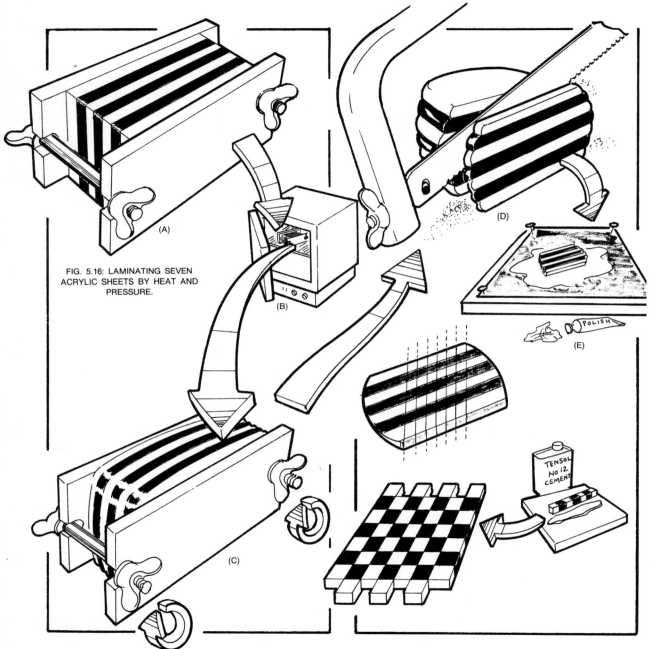

FIG. 5.16: LAMINATING SEVEN ACRYLIC SHEETS BY HEAT AND PRESSURE.

(A)

(B)

(C)

(D)

(E)

POLISH

TENSOL NO 12 CEMENT

Lamination in the oven, using heat and pressure

As an illustration of this method, suppose that you wish to laminate seven coloured acrylic sheets to make jewellery. Cut them into blanks 60 mm × 40 mm. Wash them and clamp them firmly together in the required arrangement between metal plates slotted for clamping nuts and bolts (Figure 5.16A). Place the assembly in the oven at 165°C (Figure 5.16B). The heat will penetrate slowly through the assembly. After 15 minutes, remove the unit from the oven and tighten up the bolts (Figure 5.16C). Put the unit back immediately for further heating and increase the temperature to 170°C. After a further 10 minutes at this temperature, remove the unit again and tighten the bolts even further. Then return the assembly to the oven for a further 10 minutes at 170°C. Finally, remove the unit from the oven, tighten up a little more, and leave the closed assembly to cool. When it is cold, unbolt it and you will find the plastics layers fully fused together.

The laminations may now be cut up (Figure 5.16D) and rearranged as required. The best tool for cutting is an 18 tpi (teeth-per-inch (25 mm)) high-speed hacksaw blade, which you can use to obtain thin slices of laminations suitable for decorative jewellery work.

Each slice may be filed flat and rubbed down with various grades of wet-and-dry paper (used wet) (Figure 5.16E). Cutting paste, metal polish and perspex polishes nos. 1 and 2a may be used to bring up a high-gloss finish.

Using bag welders

Bag welders are available in two main types:
 (a) solid-element (heat-sealers);
 (b) ribbon-element (impulse welders).
Both types have two straight jaws, one movable and the other fixed. They look very similar and,

in both, heat from the elements melts two thin plastics film layers under pressure. The pressure is applied until the materials have cooled.

Solid-element heat sealers suitable for welding such items as freezer bags are available with one heater element. More sophisticated versions are made for industrial use that have two electrical elements. In both cases the element comprises an electrical resistor core that heats a metal bar. Because of its mass, the bar takes time to warm up and cool down. To prevent the plastics bag sticking to the metal, a PTFE-coated foil ribbon covers each element.

In the single-element sealer, the plastics film is pressed against a silicone rubber pad; in the double-element machine the plastics film is sandwiched between the two elements. The heaters are temperature-controlled, and many machines have timers so that good-quality welds can be repeated.

Ribbon-element impulse welders work in a different way. Both of the jaws are fitted with a ribbon element. Each element is a thin metal ribbon that heats up and cools down very quickly, the quality of weld being controlled by a timer. The plastics films are placed between the elements, the jaws are closed, and the material is heated under pressure for a specific amount of time. The heater is switched off, but the pressure is maintained until the joint has cooled. The jaws are opened and the joined film is removed as one piece.

Some impulse welder machines are fitted with PTFE-coated foil. This is not really necessary because the metal ribbon element cools so quickly that it releases itself from the plastics materials.

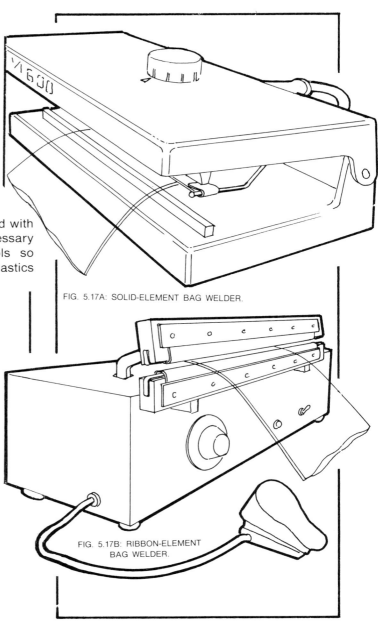

FIG. 5.17A: SOLID-ELEMENT BAG WELDER.

FIG. 5.17B: RIBBON-ELEMENT BAG WELDER.

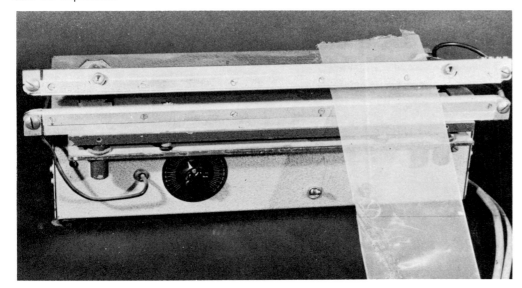

A bag welder used for making polyethylene bags from continuous lengths of 'layflat' tubing.

53

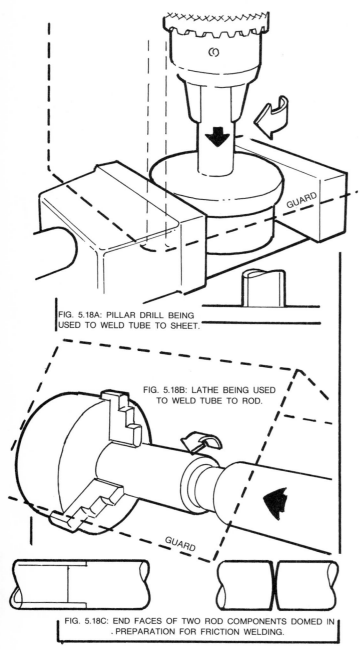

FIG. 5.18A: PILLAR DRILL BEING USED TO WELD TUBE TO SHEET.

FIG. 5.18B: LATHE BEING USED TO WELD TUBE TO ROD.

FIG. 5.18C: END FACES OF TWO ROD COMPONENTS DOMED IN PREPARATION FOR FRICTION WELDING.

Spin friction welding

When two rigid plastics surfaces are rubbed together quickly under pressure, heat is generated. This heat is sufficient to cause both surfaces to melt and mix. When the rubbing is stopped the materials will weld if they are held firmly together, and only a thin joint line will be visible. The welded joint will be as strong as the component plastics materials.

The process can be carried out on a *pillar drill* (Figure 5.18A), with one component moving and the other remaining stationary. For example, a piece of tube can be joined to a piece of sheet material. It is essential for both pieces of material to be held very firmly — the tube in the chuck and the sheet on the drill bed. The process can also be carried out on a *lathe* (Figure 5.18B), especially where two pieces of tube or rod must locate accurately, centre to centre. In this case one component is held in the chuck and the other is mounted in a holder on the tailstock.

In both these cases, the machine is switched on for about 25 seconds with the joint line firmly in contact. After being switched off, the machine will continue to run for a while. The pressure should be maintained for another 20 seconds after the machine has stopped to give the material in the joint time to solidify. The pressure only needs to be sufficient to keep the materials in contact with one another.

When you are joining two pieces of round bar (Figure 5.18C) it is advisable to dome the ends before welding. This guarantees that the central parts will weld and allow the surfaces to come together progressively. The inner and outer surfaces are moving at different speeds, which will cause uneven heating over the surfaces.

This process is used commercially for joining round, rigid injection-moulded components. Manufacturers usually make special spring-loaded jigs in which to hold the components. Sometimes the components revolve in opposite directions so that they need to be in contact for only a few moments. Often both parts finish up spinning together in the same direction at the same speed; this saves time, since, when the assembly stops the weld is already sufficiently rigid for the parts to be removed from the jig.

Warning: This operation should be carried out under supervision in a well ventilated room because fumes are given off. Eye protection must be worn because some materials splinter.

High Frequency (RF) welding

Radio frequency welding is used to join and emboss plastics film materials that have high electrical resistance. The machine creates an electrical field through the materials (usually PVC, polyurethane, or cellulose acetate) causing molecular vibration between them at the join. Momentarily the junction reaches melt temperature and under pressure from a die fuses together. Time can be adjusted to suit materials and thicknesses. Wallets, book covers, paddling pools and rainwear items are made in the way.

Ultrasonic welding

These industrial machines convert high-voltage electrical energy into high-frequency mechanical vibrations (20 KHz) to join either rigid thermoplastics parts or films and fabrics. Local frictional heat develops where the parts rub together on the join line fusing the parts together under pressure. This is a very fast, clean and strong method of welding that does not stress the moulding.

PROJECTS AND ACTIVITIES

(1) Design and make a device to test the strength of a cemented joint. Prepare several joint samples at the same time under the same conditions. Test the strength of the joints over several days, taking one sample every 24 hours. Does time have any effect on the strength of the joint?

(2) Prepare two pieces of acrylic tube about 25 mm in diameter by cutting them into 25 mm lengths and rubbing the ends down until they are smooth and parallel. Put one tube into an oven, heat it to 90°C, then switch the oven off and leave the tube to cool in the oven overnight. Cement two end-pieces onto each tube and leave them for two days. What has happened to each tube at the end of this time?

(3) Design a young child's inflatable toy that can be made with a rotary or plate welder. Using polyethylene sheet, make a full-size working model.

(4) Using the very fine polypropylene tissue that a butcher uses to wrap meat, design and make a kite. More advanced students can undertake the same problem and then take an aerial picture of their home using a cheap plastic camera.

(5) Most people buy a monthly magazine. Design and make, using card and film PVC, a folder to hold 12 issues.

(6) Using polyethylene sheet, design and make an inflatable tent that can be erected to full size and kept inflated using a vacuum cleaner (on blow).

(7) Many people lack confidence in the swimming pool and never really learn to swim. Waterwings and inflatable arm bands are an insult to teenagers. Can you design a device that exploits the waterwing principle but is fun to use — even for a good swimmer? In this way a poor swimmer would be able to use it without looking out of place.

(8) When using direct heat welding techniques why is it necessary to use clear cooking foil between the welder and the plastics material?

(9) Next time you pass the local gasboard maintenance engineers laying new gas mains in the street, ask the foreman to show you how the gas pipes are joined together. Find out what materials are used and prepare an illustrated chart for the plastics area of your school CDT room.

(10) What materials are unsuitable for cementing? Find some samples and show how they can be joined.

(11) Crazing often occurs after mouldings have been joined with solvent or solvent based cements. Prepare a sample to show this effect (use a clear moulding) and then prepare another sample with full explanation showing a good joint of similar type free from crazing. Describe what has happened and how you have prevented this occurring in the second sample.

(12) Find examples of the following jointing techniques used on household products:
   (1) mechanical fixings — plastic rivets, plastic screw threaded components, plastic push fittings;
   (2) cemented joint;
   (3) welded joint in flexible sheet;
   (4) welded joint on extruded or rigid moulded product;
   (5) double sided adhesive foam or tape joint.
   Examine each closely and suggest why each technique is appropriate to the application.

(13) Write to a manufacturer of household cements and adhesives and ask for two charts showing their products and applications. Give one to your teacher for display purposes and clip the other in your plastics file for future reference. Next time you have to join different materials consult the chart to find out the best adhesive material to use. Find out the difference between an adhesive and a cohesive.

(14) If you have the use of a hot air welder and welding rod prepare a chart showing good and poor welds. Some of the samples should be sectioned and some should show both the raw weld and the finished surface after cleaning back. Use PVC material and identify the speed and temperature used to make the sample.

(15) What treatment should be carried out to reduce the risk of crazing on a clear acrylic sheet bubble blow moulding before additional pieces of acrylic block are cemented to it?

(16) Carry out some experiments to find a suitable dye or colouring agent to make clear Tensol No. 12 cement coloured. Then design a pair of earrings that are made from small pieces of clear $1\frac{1}{2}$ mm thick acrylic sheet laminated together with coloured cement.

(17) Make a laminated car key fob from two pieces of 3 mm thick acrylic sheet. The two pieces of acrylic sheet should be heated as instructed on page 52. Try trapping a stamp, photograph or fishing fly between clear sheets for decorative effect. Record what you do both diagramatically and in words. Suggest other uses for this process to your teacher for younger children to do.

6 INJECTION MOULDING

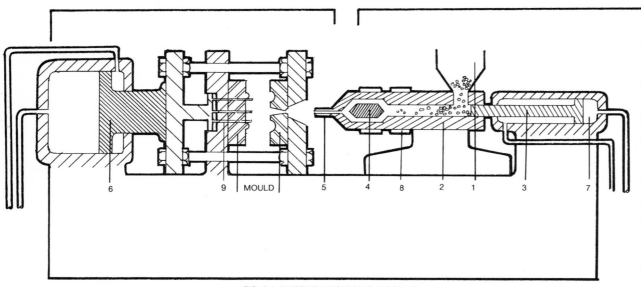

FIG. 6.1: INJECTION-MOULDING MACHINE.

1: HOPPER	4: TORPEDO SPREADER	7: RAM-DRIVING SYSTEM
2: BARREL	5: NOZZLE	8: HEATER ELEMENT
3: RAM	6: MOULD-CLAMPING SYSTEM	9: EJECTOR PINS

Injection moulding is a high-speed process in which powdered or granulated thermoplastics are heated, melted and forced under pressure into a mould. The material in the mould then cools, forming a component that takes the shape of the mould cavity. New moulds are fitted to the machine whenever a new type of product is required. Each mould can be used to produce many thousands of mouldings. If, for example, you were asked to make 10,000 Lego bricks, the best way to make them would be by injection moulding.

Injection-moulding machines
Figure 6.1 shows the main parts of an injection-moulding machine:
(1) a *hopper* for the cold plastics powder or granules;
(2) a heated *barrel*;
(3) a *piston ram* or *screw ram*;
(4) a *torpedo* or spreader unit;
(5) a *nozzle*;
(6) a *mould-clamping system* (for opening and closing);
(7) *ram-driving system*;
(8) *heater* elements.
(9) *pins* to eject moulding.

Working procedure
Plastics powder or granules are fed from the hopper into a steel barrel. The barrel is a hollow tube fitted with electrical heaters on the outside at one end, and with the piston ram or screw ram at the other. An opening in the top of the barrel allows plastics granules to fall into the internal barrel chamber at the front of the ram. When loaded, the ram forces that material to move forward and compresses it. The heaters on the outside of the barrel are at different temperatures, from cool at the hopper end to plastic 'melt' temperature at the nozzle end. This controls the steady increase of plasticity in the material.

When the ram moves forward to the injection position it brings the nozzle directly into contact with the *sprue* (entry point) into the mould. As the ram compresses the plastics material it pushes the complete barrel system forward to make a tight seal between the nozzle and the mould. The hot plastics material flows from the barrel, through the nozzle and into the mould.

An external view of a commercial injection moulding machine.

The mould

In the mould the material flows (under pressure) along the main sprue passageway to the cavity, which is a large space in the shape of the article to be made. Moulds can contain several cavities: when this is so, each cavity is connected to the sprue passage by channels known as *runners*. The mould is made of steel in two parts, which are held firmly together for the period of injection and cooling. One half of the mould is fixed while the other half moves with the rest of the machine. The two surfaces where the mould halves meet are flat and often create a line around the moulding called the *split line*. When the mould is made, the runner channels are machined into one or both of these surfaces. After the hot plastics has been injected, it fills the cavity, the runners and the sprue passage. It is allowed to cool for a short period known as *dwell time*. During the dwell time the plastics material changes from a syrupy fluid into a solid. Then the mould opens, the moulding, runners and sprue remaining on the fixed side. This clears the sprue passage, runners and cavity on the movable half. The moulding is then pushed out of the fixed part of the mould by *ejector pins*, leaving the mould interior clean.

As the mould opens, the ram and the barrel assembly move back to their starting positions. When the ram passes beyond the hopper opening, fresh plastics material falls into the barrel. The machine is ready to make a new moulding.

The cycle

This whole operating procedure is known as a *cycle* (Figure 6.2). The amount of material used in one cycle is known as a *shot*. Such machines are often rated by the maximum shot weight of general-purpose polystyrene plastics material that can be injected in one cycle.

The principle behind all the machines is com-

paratively simple: controlled heat; controlled pressure; a repeatable cycle.

Commercial machines differ from school hand-operated machines in having screw ram systems. The simplest screw ram consists of a single screw that is made to rotate for a short period (Figure 6.3). This action of the screw moves the plastics materials along the barrel and helps to plasticize (melt) the material by the 'shearing' action of the helix against the barrel wall. After a short time the screw ram stops revolving and is forced forward, then acting like a piston.

View of a commercial injection mould showing the positive parts of the mould, with ejector pins projecting.

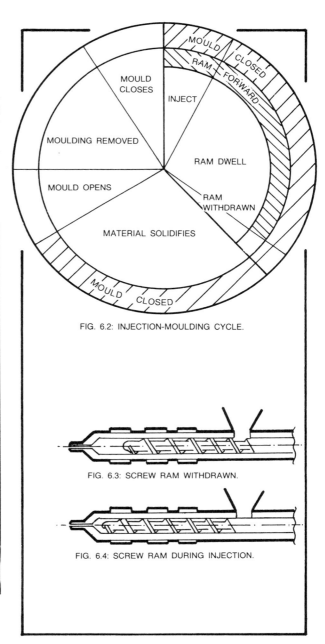

FIG. 6.2: INJECTION-MOULDING CYCLE.

FIG. 6.3: SCREW RAM WITHDRAWN.

FIG. 6.4: SCREW RAM DURING INJECTION.

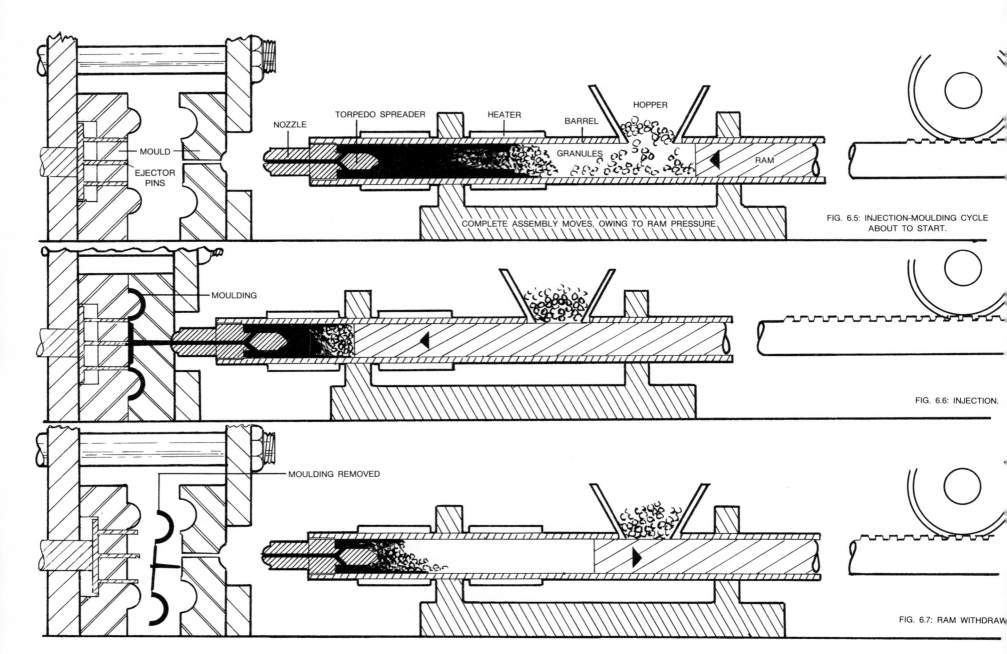

MOULD

EJECTOR PINS

NOZZLE

TORPEDO SPREADER

HEATER

BARREL

HOPPER

GRANULES

RAM

COMPLETE ASSEMBLY MOVES, OWING TO RAM PRESSURE

FIG. 6.5: INJECTION-MOULDING CYCLE ABOUT TO START.

MOULDING

FIG. 6.6: INJECTION.

MOULDING REMOVED

FIG. 6.7: RAM WITHDRAW

58

THE HORIZONTAL AND VERTICAL INJECTION-MOULDING MACHINE

Injection-moulding machines are designed to operate either horizontally or vertically. Figure 6.5–6.7 show the main movements that take place in a horizontal machine to produce a moulding. Figure 6.5 shows the machine at rest, with the mould open and the ram drawn back to allow material to enter the barrel from the hopper.

Figure 6.6 shows the mould halves closed, the barrel assembly moved up to make the nozzle press firmly against the mould sprue entry point, and the material flowing through into the mould cavity. After a short dwell time in this position while the plastics freeze in the mould, the ram withdraws and the whole assembly moves back.

In Figure 6.7 the mould opens, and the rigid (though still hot) plastics moulding is ejected by ejector pins on the fixed mould side. It is important to remember that the moulding is deliberately *held* on this side so that the sprue runner system is cleared. Look closely at Figure 6.14 (page 63): the cold slug is shaped to stay in the mould half that is opposite to the sprue. The cold slug is said to be *undercut* and ejects only because a large ejector pin is positioned at this point. Notice that all the ejector pins are mounted on this mould half. As the ram continues to move back past the hopper opening, fresh granules drop into the barrel. At the mould end the moulding drops clear and the cycle starts again.

Figure 6.9 shows a typical vertical ram machine of the type often used in schools. Notice that the ram operates *through* a small barrel hopper. Usually measured quantities of granules are dropped into this hopper through a chute from a second hopper placed above the machine.

The ram is pushed down, forcing material down into the heater barrel. The pressure on the ram is applied by a rack-and-pinion mechanism (similar to that used in a pedestal drilling machine), or by compressed air. Commercial ram machines use hydraulics (oil pressure) to drive the ram.

As the ram applies pressure to the granules, the whole assembly moves down guide rails until the nozzle makes firm contact with the mould. The mould is held firmly shut by two screw clamps mounted opposite one another. These have different threads to permit coarse and fine adjustment, which is important for accurate positioning of the mould below the nozzle. The sprue should lie on the mould split line, which is parallel to the two clamping blocks. Figure 6.8 shows a typical mould and the moulding it would produce. Guards are fitted on these machines to prevent hot plastics material from squirting from the machine onto the operator. These guards must be used for *every* machine cycle.

After a short dwell time the ram is withdrawn and the cycle is repeated.

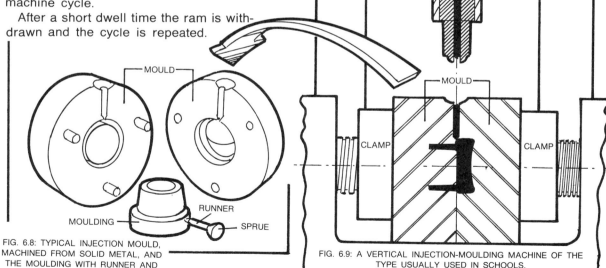

FIG. 6.8: TYPICAL INJECTION MOULD, MACHINED FROM SOLID METAL, AND THE MOULDING WITH RUNNER AND SPRUE STILL ATTACHED.

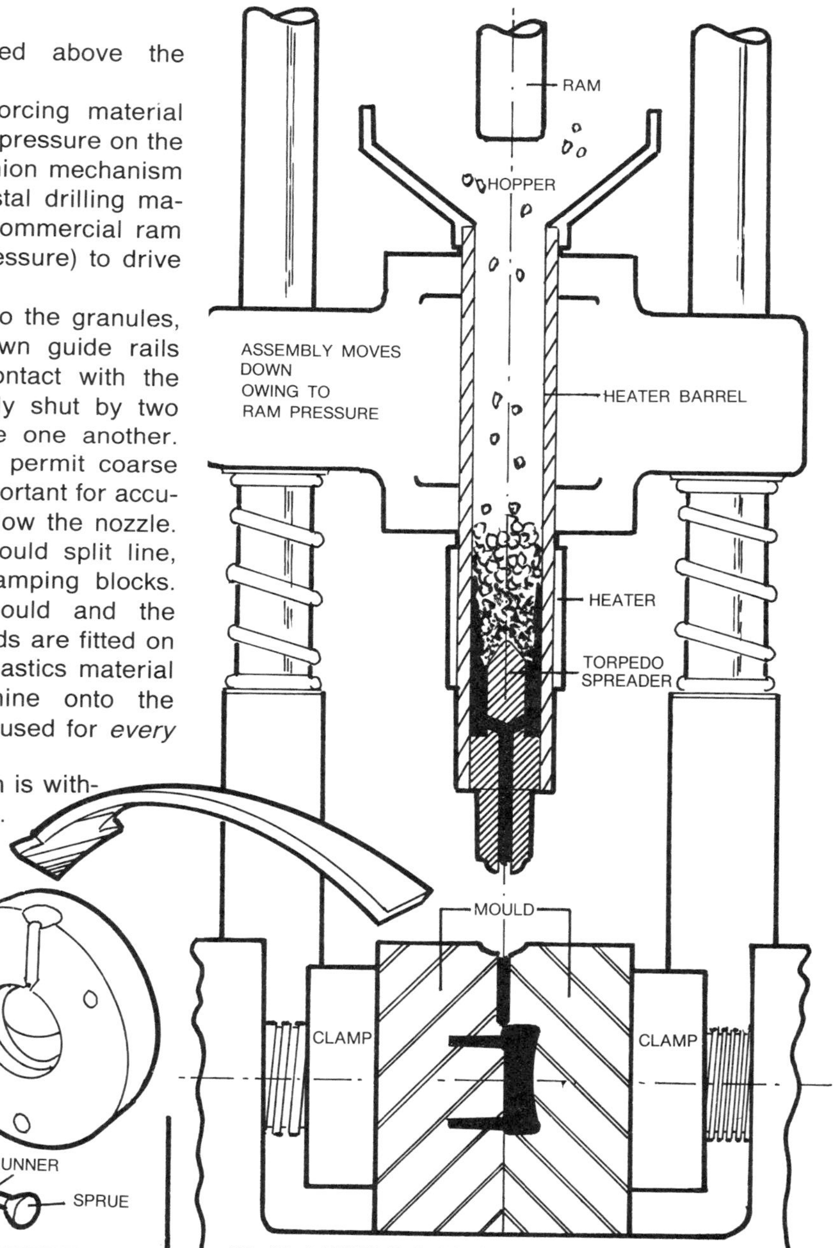

FIG. 6.9: A VERTICAL INJECTION-MOULDING MACHINE OF THE TYPE USUALLY USED IN SCHOOLS.

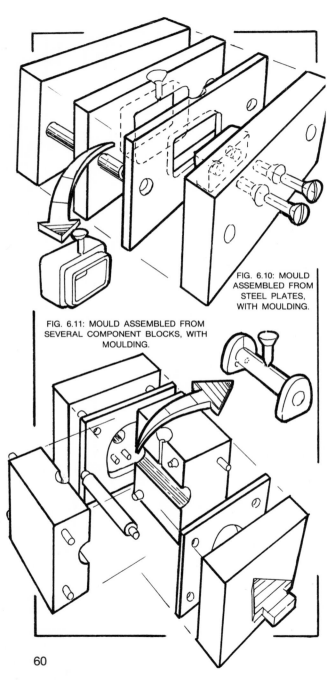

FIG. 6.10: MOULD ASSEMBLED FROM STEEL PLATES, WITH MOULDING.

FIG. 6.11: MOULD ASSEMBLED FROM SEVERAL COMPONENT BLOCKS, WITH MOULDING.

MOULD DESIGN FOR SCHOOLS

Moulds may be made in several different ways:
(1)   machined in two halves from solid metal (Figure 6.8 on page 59);
(2)   assembled from steel plates (Figure 6.10);
(3)   assembled from component block parts for each half (Figure 6.11);
(4)   cast in type-casting metal in two halves (Figure 6.12);
(5)   cast in two halves from metal-filled epoxy resin (Figure 6.13).

When selecting mould material, check that the two halves when assembled will fit the machine and if possible arrange the split line to coincide with the nozzle position. The screw clamp arrangement should be strong enough to keep the mould halves firmly closed.

Once these essentials have been established, the next major point is to ensure that the planned mould cavity does not encroach on the guide pin locations or get too close to the mould exterior wall. About 12 mm of material should surround the cavity. All moulds need to be tapered so that the finished mouldings can be removed; however, it is possible to keep this to a minimum. The terms 'taper', 'draft angle', and 'angle of withdrawal' all mean the same thing and are discussed in detail in Chapter 1.

Warning: Eye protection and heat resistant gloves must be worn to handle hot moulds and mouldings. Moulds must not fracture and guards must be in place during moulding. Never touch molten plastics materials.

Examine a plastics model aeroplane kit closely and you may see the sprue, ejector pin marks, and undercut cold slug impressions, as well as mould cavity numbers, cavity shapes, gates, runners and mould split lines (Figure 6.14). The expert eye can 'read' a moulding and can learn much about the mould.

It is essential that the mould halves match up exactly, and the first operation to ensure accurate matching is to drill the holes for the guide pins, whose centres may then act as datum points for measurement.

It is sometimes possible to mix the five typical methods of construction, but it is more usual to employ one method for each mould. When metal-filled epoxy resin is used with steel or brass filler powders, there may be problems resulting from different expansion rates with increases in temperature.

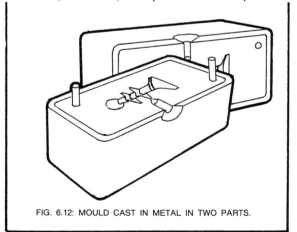

FIG. 6.12: MOULD CAST IN METAL IN TWO PARTS.

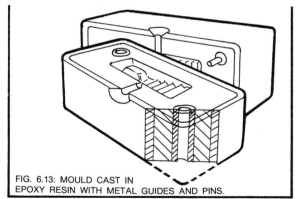

FIG. 6.13: MOULD CAST IN EPOXY RESIN WITH METAL GUIDES AND PINS.

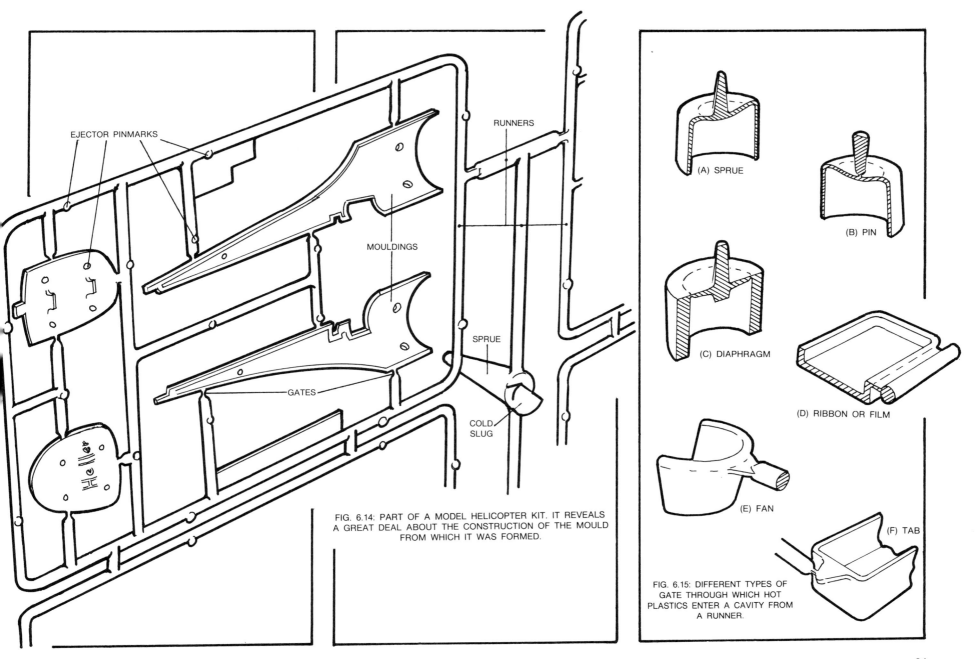

EJECTOR PINMARKS

RUNNERS

MOULDINGS

SPRUE

GATES

COLD
SLUG

FIG. 6.14: PART OF A MODEL HELICOPTER KIT. IT REVEALS
A GREAT DEAL ABOUT THE CONSTRUCTION OF THE MOULD
FROM WHICH IT WAS FORMED.

(A) SPRUE

(B) PIN

(C) DIAPHRAGM

(D) RIBBON OR FILM

(E) FAN

(F) TAB

FIG. 6.15: DIFFERENT TYPES OF
GATE THROUGH WHICH HOT
PLASTICS ENTER A CAVITY FROM
A RUNNER.

Sprues, runners and gates

The sprue entry, through which the plastics material flows into the mould, must be 'countersunk' (given a bevelled edge) so that the injection nozzle can seat accurately. It is important that the nozzle opening should line up accurately with the sprue opening, in a way that prevents material from escaping sideways. The sprue channel should be tapered if it does not lie on the split line, so that the moulded sprue can be separated from the mould. The sprue can feed either a runner system (Figure 6.14) or the mould cavity directly through one of several different types of *gate*.

When the sprue feeds a runner system, a small cavity called a *cold slug* is provided. The mould is cooler than the material being moulded, which causes the first of the material to cool as it flows down the sprue. This front of cold material flows across the runner passage into the cold slug cavity, leaving the way clear for the hotter material following behind to flow sideways down the runner. On complex commercial moulds there is sometimes more than one cold slug cavity along a runner before the material eventually passes through the gate into the mould cavity.

Runners can be round or trapezoidal (trapezium-shaped) in section. A round runner will lie on the split line, half being in each side of the mould. A trapezoidal runner is usually a channel cut in one mould half, facing up to a flat surface on the other half. On large moulds runners will be up to 6 mm in diameter, the outer layers chilling while molten materials continue to flow through the core.

Different types of gate are shown in Figure 6.15. They vary in design to suit the product and the plastics material. One important feature is common to most types of gate: it is narrower than the runner, often being a very fine slit or small hole. This serves two purposes; (1) it does not leave an unsightly mark on the moulding; (2) it causes frictional heat to be generated as the material squirts under considerable pressure into the cavity. Reheated in this way, the material stays molten and flows to all parts of the cavity. Because of the pressures involved, the air in the system is forced out through the split line, along the sides of ejector pins and along mould component joint lines.

Ribs, bosses and corners

Injection-moulded flat surfaces are liable to bow inwards. To prevent this, mouldings are often ribbed on their reverse (unseen) surfaces to give stiffness (Figure 6.16). A rib should be tapered

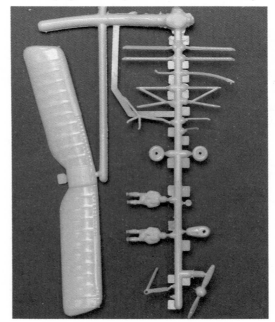

Part of a model aeroplane kit showing a moulded frame that acts as the runner system to carry hot plastic to the mould cavities during injection moulding.

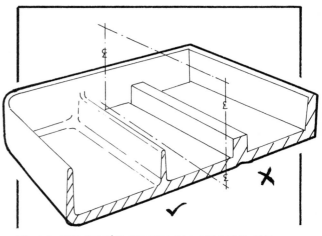

FIG. 6.16: TWO DESIGNS FOR RIBS IN A MOULDING. THE SLENDER, TAPERED RIB, RADIUSED WHERE IT MEETS THE FLAT SURFACE, IS BETTER. THE THICK, SQUARE RIB WILL PROBABLY CAUSE A SINK MARK BECAUSE OF UNEVEN COOLING AND SHRINKING, AND WILL BE DIFFICULT TO SEPARATE FROM THE MOULD.

with about 2° draft angle on both sides and should have about $\frac{2}{3}$ the thickness of the main surface. Where the rib joins the main moulding the corner must be radiused. These precautions will prevent *sink marks* from appearing on the good surface of the main moulding. Sink marks are caused by the material on the inside of the moulded section cooling and shrinking more slowly than the outer skin. Short fat ribs with sharp corners create noticeable sink marks. Locating bosses create similar problems and should be moulded near corners where possible. Their height should be less than twice their diameter.

Where two sides, or one side and the bottom, of a plastics moulding meet at a corner, the shape should be radiused so that the thickness is constant (Figure 2.9, page 21). Sharp corners are not as strong, and can show sink marks caused by the difference in thickness at the corner.

INJECTION MOULDING IN SCHOOLS

Many schools have designed and made their own moulds and in some cases their own injection-moulding machines.

The machines available for school use are of the simple piston-ram type, and the mould halves are usually clamped together using screw-clamping techniques. The ram system is often operated by hand through a rack-and-pinion drive, but the alternative of a compressed-air drive is available for some machines. A single collar heater of about 250 watts rating is all that is necessary to heat the short barrel on this type of machine.

These machines are suitable for moulding general-purpose and high-impact polystyrene, polyethylene, and polypropylene. Polyvinylchloride should not be used, because it may break down, releasing chlorinated gases that can attack the heater barrel and moulds. Although other thermoplastics materials may be available, it is wise to check whether they are suitable for piston-ram machines. When purchasing plastics for injection moulding check whether they are hygroscopic (absorb water from the atmosphere), and whether they 'gas' easily (burn and break down in the heater chamber). Materials that have these properties should not be used.

It is, of course, possible to cut up polyethylene bottles, grind the material into small pieces in an old hand mincer and reuse them to make a new product.

A low-cost method of producing small injection mouldings

A number of very good electrical glue guns, using glue sticks (pellets of thermoplastics material), can be bought cheaply today. Such a gun has a heater chamber with a nozzle at one

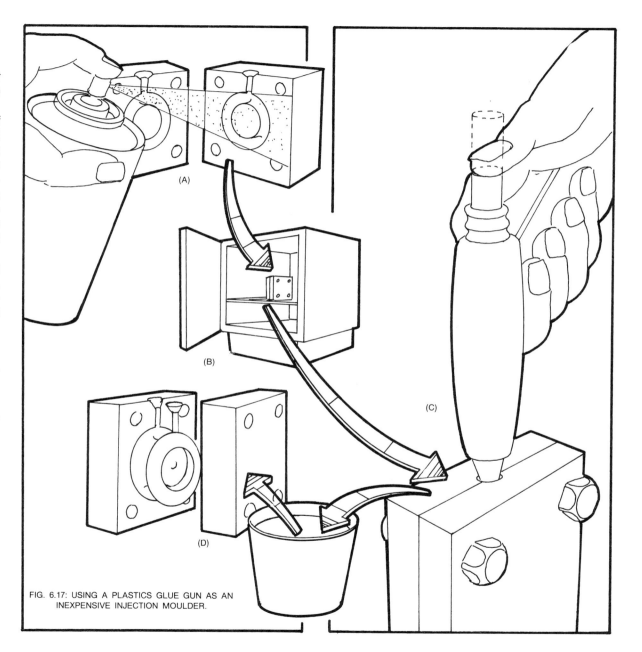

FIG. 6.17: USING A PLASTICS GLUE GUN AS AN INEXPENSIVE INJECTION MOULDER.

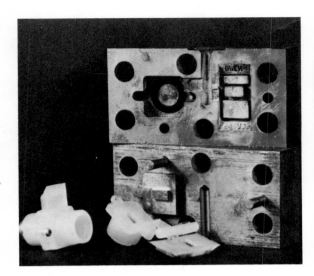

Injection mould made by a student for the construction of a valve.

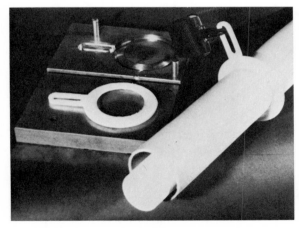

A container for carrying rolled drawings on a bicycle. The design centres around a flexible polypropylene ring made by injection moulding used to secure a hanging shoulder strap to the ABS extruded tubular container.

Injection mould made by a student, for a special clip to hold a medical tube near a baby's mouth.

Injection mould made by a student for the moulding of a 'pill box'.

An example of how a metal component can be moulded-in during the injection moulding process.

end and an opening at the other, into which you press the plastics pellet with your thumb (Figure 6.17).

The chamber is heated electrically. Pushing in a new pellet forces the material already in the heated chamber out through the nozzle — normally into a joint between pieces of wood.

However, you can press the gun nozzle against the sprue opening of a small injection mould, forcing the hot plastic glue material into the cavity. All parts of the mould must previously have been sprayed with a silicone release agent *in a fume cupboard* (Figure 6.17A).

The sticks of plastics polymer are tough and flexible, giving small mouldings with the same qualities. The material holds its heat for a relatively long time when cast into a mould. The mould however must be heated (Figure 6.17B) to keep the material molten while it flows into the cavity (Figure 6.17C). This high mould temperature (90°C) is necessary because the pressure is low. The mould must then be cooled before it is opened (Figure 6.17D). Heat-resistant leather gloves are essential to handle the hot mould.

Mouldings should be designed to have thin (2 mm) walls because the polymer glue material shrinks in thick sections.

For schools that do not have injection-moulding machines this can be a cheap method of obtaining satisfactory small injection mouldings. A glue gun costs under £15 and the pellets cost a few pence each. Both are available from hardware and DIY stores.

The glue stick pellets are usually made from EVA (ethylene vinyl acetate).

A further example of a metal component moulded-in during the injection moulding process.

Injection mould made by a student to create end plugs for a tubular container.

PROJECTS AND ACTIVITIES

(1) Design a device, suitable for manufacture by injection moulding, that can be mounted in a shower cabinet to dispense a measured quantity of liquid soap. It should be possible to see the level of soap, and the soap must not drip onto the floor. Make a model of your design in cardboard or plastics sheet material.

(2) Design a constructional toy for nursery children, based on plastic drinking straws and an injection-moulded joint unit. Make a detailed drawing showing how the joint unit works, and make a model in the most suitable material. Prepare a drawing showing the essential features of the mould from which the joint unit could be made.

(3) Design a support for a plastics glue gun that will make it more effective as an injection moulder than it is when unsupported. A simple ram system could also be incorporated.

(4) Take two small identical polystyrene articles (such as a pair of clothes pegs) and heat them to their softening temperature. They should change shape. Examine their new shape after they have cooled, describe how they have deformed and suggest why they have taken up similar forms.

(5) Have a competition with your friends to see who can find the largest injection moulding, at home and at school. If possible measure its size, weight and thickness, and identify the material used.

(6) Examine five injection-moulded plastics products in your kitchen and draw them, paying particular attention to the following details, which should also be sketched:
   (1) the gate, or point where the material was injected;
   (2) any split lines;
   (3) ejector pin marks;
   (4) sink marks;
   (5) any moulded-in screw fittings;
   (6) a cross-section of the material.

(7) Choose an article (such as a clothes peg or a pen) for which three different designs, in different materials, made by different processes, are available. Describe how the materials and processes have contributed to the qualities of the product.

(8) Take a cheap plastics product made by injection moulding and cut it up into sections. Measure the material thickness in the main body and in ribs and bosses. Check for sink marks at intersections and square corners. See if the ribs have been used as a runner system during moulding to help the distribution of material.

(9) Find an unwanted domestic electrical appliance (for example, an old hair dryer) whose outer casing has been injection-moulded. Take it to pieces and then prepare an 'exploded' drawing to show how it has been assembled. Weigh the individual mouldings and find out what material has been used. Through your local library or a local plastics firm, find out the cost of one tonne of the material. Then calculate the material cost of each plastics component and the number of components that can be made from one tonne.

(10) See if you can find an injection-moulded product that has been 'blown' — in other words, had gas introduced into the melt, turning it into a foam. Examine it closely and describe how it differs from similar products made by conventional injection moulding.

7 EXTRUSION

Extrusion is the process used to make continuous lengths of material with a constant end section, such as curtain rails, plastic pipes, etc. This process can produce:

(1) special sections ('U', 'L', 'T', 'H', 'E'– Figure 7.1);

(2) tube (Figure 7.2);
(3) wire coatings (Figure 7.3);
(4) film (Figure 7.4);
(5) extrusion blow moulding — mainly bottles (Figure 7.5);
(6) sheet (Figures 7.7 and 7.8).

The choice of material will affect the physical properties of a plastics extrusion, making it either rigid or flexible, clear or coloured, suitable for exterior use or particularly resistant to chemical attack. Granulated materials can be specially prepared for extrusion, although in many instances the materials used for injection moulding are also satisfactory. Materials are chosen on the basis of their properties and suitability for particular applications. For example, a material with special chemical resistance would be chosen for pipework in a laboratory.

THE EXTRUSION MACHINE

Figures 7.1–7.7 show several different types of extrusion machine and dies. These machines are all similar in principle — they have a screw, revolving continuously inside a heated barrel, and material in granulated form is fed from a hopper to the screw, which carries it forward into the heated barrel. The material changes from granules into a viscous liquid through the action of heat and the shearing forces of the screw against the barrel wall.

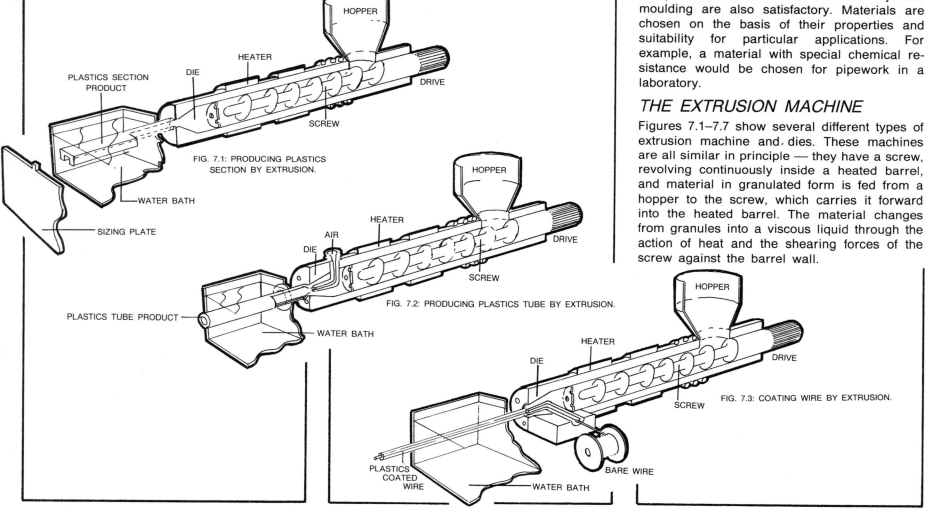

FIG. 7.1: PRODUCING PLASTICS SECTION BY EXTRUSION.

FIG. 7.2: PRODUCING PLASTICS TUBE BY EXTRUSION.

FIG. 7.3: COATING WIRE BY EXTRUSION.

The *die* is fitted to the end of the barrel (Figure 7.6) and it is the shape of the die that determines the profile of the product that emerges from the machine. Between the die and the end of the screw there is usually a *breaker plate*, which prevents the material from revolving further after leaving the screw and screens out any foreign bodies. The hopper end of the barrel is cool and gradually increases in temperature towards the die, so that the material is in a fully molten state as it passes through the die. The screw creates a considerable pressure on the material as it forces it along the barrel, and the pressure is increased as the material reaches the conical shape of the small die opening.

Immediately the material has passed through the die it expands, an effect known as *die swell* (Figure 7.6). The shaped material has to be drawn away at a constant speed. The plastics product as it leaves the extrusion machine is called an *extrudate*. The plastic extrudates for rod, tube, sections, and wire-coating profiles all pass through a water bath for cooling. Additional formers are often situated in the water bath to help the material keep its shape. After leaving the water bath the extrudate passes between caterpillar tracks (known as *haul-off gear*) before being cut into lengths or coiled.

When wire is coated with plastics materials the wire passes into the extrusion machine from a bobbin drum, collects a thin coating of plastic as it passes through the die and is then cooled in a waterbath before being coiled (Figure 7.3).

To make plastic tube, film and bottles (Figures 7.2, 7.4 and 7.5) it is necessary to prevent the material collapsing in on itself as it leaves the die. Compressed air is introduced into the extrusion machine through a pipe, in a way similar to that used for coating wire. In this way the inside wall of the extruded tube is cooled and prevented from collapsing. With film the plastic material inflates into a long sausage-shaped bubble, which is closed at the end where it is coiled up (Figure 7.4). Only enough compressed air is blown into the tube to keep it in a constant state of inflation, which determines the film thickness. Later one edge of this 'lay flat tube' is slit to leave a long roll of film folded along its length.

Bottles are produced by a post-forming operation on tube (Figure 7.5). A short length of tube, still very hot, leaves the die face and is collected between the two halves of a split mould. This piece of hot tube extrudate is called a *parison*. The parison is pinched between the mould faces at one end, closing the tube to form the base of a bottle. The other end fits over a compressed air pipe and air is immediately blown into the parison. The parison inflates to fill the bottle-shaped cavity in the mould halves.

Cut a 'squeezy bottle' in half along its main axis. You will immediately be able to see the original parison tube where it has been squashed and the thin wall section where it has been inflated. On the outside you can see the split line of the mould. You may also find some indentations on the bottom — these positioned the bottle on another machine for printing.

Plastics sheet is also made using the extrusion process. Polystyrene, ABS, polyethylene, polypropylene and CAB are common materials extruded into sheet and then later vacuum-formed into products by other manufacturers.

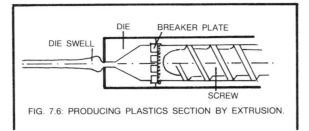

FIG. 7.6: PRODUCING PLASTICS SECTION BY EXTRUSION.

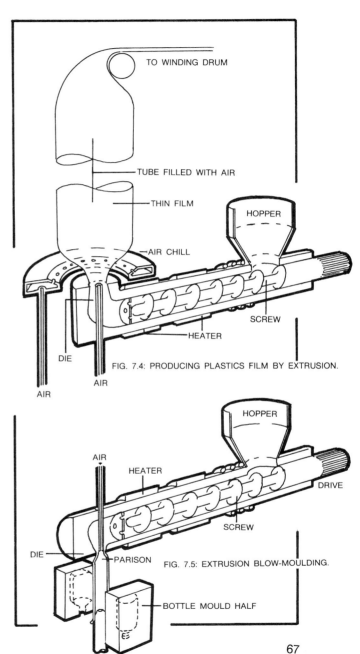

FIG. 7.4: PRODUCING PLASTICS FILM BY EXTRUSION.

FIG. 7.5: EXTRUSION BLOW-MOULDING.

67

The machine (Figure 7.7) has to be capable of heating a large volume of material continuously. Once it passes the breaker plate, it enters a *fantail die*. The die (Figure 7.8) is designed in a way that forces the hot plastics material to fan out into a flat sheet. It is usually heated and has an internal adjustable dam to restrict the flow and cause the sideways spreading to take place. When the material leaves the long narrow slot in the die it generally passes on to a pair of cold rollers that are closely spaced to control the

sheet thickness. A conveyor belt then takes the material to a circular saw, which automatically cuts across the sheet as the sheet is moving forward. The sheet drops on to a stack which is packed and despatched. The rollers can be used to create a pattern in the surface of the sheet.

In recent years plastics technology has been developed to enable all these processes to produce *foam extrudates*. A gas, usually carbon dioxide (CO_2), is forced into the molten plastics material in the barrel. It remains in the melt until the hot material leaves the die, when it expands. The surface skin has the same appearance as a normal extrusion but when cut through one can see many small gas pockets — often looking like woodgrain. Can you find any examples of foam extrusions in your home?

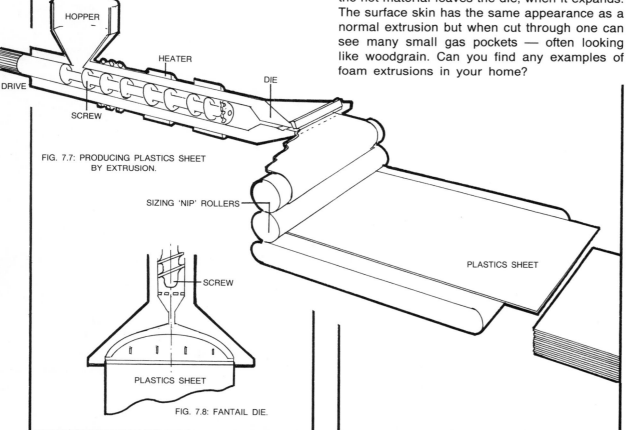

FIG. 7.7: PRODUCING PLASTICS SHEET BY EXTRUSION.

FIG. 7.8: FANTAIL DIE.

Calendering

Calendering is another process used to produce plastic film, sheet, coated fabrics and other coated materials. The main materials used are flexible and rigid PVC compounds, ABS, and cellulose acetate.

The process, Figure 7.9, is largely one of rolling out a mass of premixed plastics material between large rollers to form a continuous accurately sized sheet.

The rolls are usually made of cast iron with very accurately machined, highly polished surfaces. Each roll has internal passageways to carry hot water or hot oil which heats the calender roll and the plastic material as it passes over it. For a material like PVC the surface temperature of the roll will be between 160° and 220°C. The distance between pairs of calender rolls is called the *nip* and it can be adjusted to control the sheet thickness. Usually a machine has three pairs of nip roll points (one roll is sometimes shared between two others making two nips between three rolls). This enables the bulk of the material to be reduced gradually in thickness. The widest rolls produced at present are two metres wide and the thickness control ranges from .051 mm to 1.3 mm.

The material passes from the main calender rollers to cooling rollers where it can be embossed and polished or pass on to a printing stage before being rolled up.

Calendering is a faster process than extrusion and so the cost of production is lower. This process is mainly used to produce the material for PVC paddling pools, inflatable toys and shower curtains (from film plastics), luggage holders and credit cards (from sheet materials) and wallpaper and 'leather-look' upholstery fabrics (these are coated products).

PROJECTS AND ACTIVITIES

(1) Design light-weight 'bricks' that can be extrusion blow-moulded, that interlock with each other, and that can act as containers when filled, becoming, say, 2-litre bottles. This brick could then also be used in cold weather countries as an insulated house-brick for infilling wall spaces between structural supports. In transit as a container it should be possible to use the interlock to hold a stack together. Experiment with polyethylene bottles.

(2) You have an extrusion machine that will produce a continuous length of 3 mm-thick polyethylene rod at the rate of 2 metres per minute. When the material leaves the die it is hot enough to weld to itself. Design a light fitting that could be made using this material. (This was the way the Rotaflex company started.) Can you devise any other products that could be produced in a similar way?

(3) Some sun-lounger beds use light-weight PVC tube between aluminium frames as the means of providing comfortable body support. Measure the diameter of the tube, its wall thickness and length, and then calculate the weight of material used.

(4) Using the example described in (3) above, design one of the following:
   (1) a garden chair for a young child (about three years old);
   (2) a kneeler suitable for use when scrubbing floors;
   (3) a mattress for your cat or dog's box.

(5) Find an example of an extrusion blow-moulded bottle and an injection blow-moulded bottle (see photos on page 5). Cut them both in half (lengthwise) and describe what you see.

(6) Build a project file of as many different samples of extruded products as you can find. Decide on a method to categorize them and draw the typical shape of the die used to produce them.

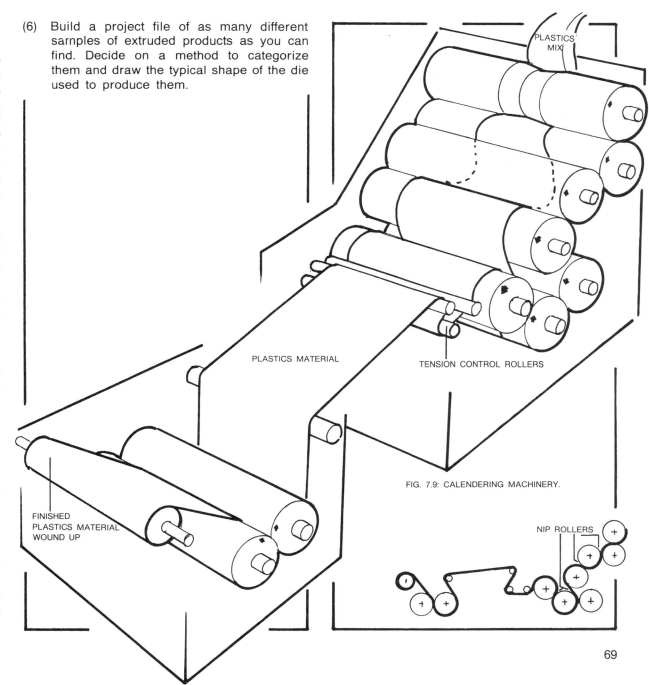

FIG. 7.9: CALENDERING MACHINERY.

PLASTICS MIX

PLASTICS MATERIAL

TENSION CONTROL ROLLERS

FINISHED PLASTICS MATERIAL WOUND UP

NIP ROLLERS

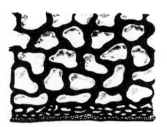

CLOSED CELL

OPEN CELL

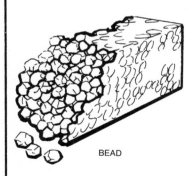

BEAD

FIG. 8.1: STRUCTURE OF CELLULAR PLASTICS.

Table 3 Expanded and foamed plastics

Structure	Type	Nature	Form	Material	Application
Closed-cell Expanded Plastics	thermo-plastic	flexible	sheet	polyethylene	plastazote, medical, life jacket buoyancy
		flexible	sheet	ethylene vinyl acetate	Evazote, packaging
		semi-rigid	sheet	polystyrene	egg cartons, packaging
		rigid	sheet	PVC	Foamex, exhibition panels and displays
		rigid	sheet	ABS	electronic casings
		rigid	block	acrylic	machined components
			injection and extrusion mouldings	most thermo-plastics	wide range of domestic leisure and office products
	thermoset	rigid	foamed in situ block	urea formal-dehyde	house insulation
		rigid	block	phenol formal-dehyde	insulation panels
		rigid	block	polyurethane	insulation panels, marine buoyancy
Open-cell Foamed Plastics	thermo-plastic	flexible	sheet	PVC	cushion flooring, leathercloth, shoes, baggage
		semi-rigid	block and some mouldings	polyurethane/elastomer	upholstery cushioning
		rigid	sheet	PVC	filters
	thermoset	rigid	block	phenol formal-dehyde	flower decoration foam (Oasis)
Bead Closed-cell Expanded Plastics	thermo-plastic	flexible	mouldings	polyethylene bead	packaging
		semi-rigid	sheet/block and mouldings	polystyrene bead	ceiling tiles and mouldings, packaging
Natural Rubber		flexible	moulded shapes		bendy toys
		flexible	block		cushions
		flexible	extruded sections		car door seals

CELLULAR MATERIALS

When a bar of Aero chocolate is broken open, you can see a large number of air bubbles surrounded by chocolate. This is a *cellular* structure. Most plastics can be produced in a similar way, with a gas being introduced into the material at some stage during its processing. *Cellular plastics* can have two types of structure:
(1) closed cell (often called *expanded plastics*);
(2) open cell (called *foamed plastics*).
Figure 8.1 shows both types. It also shows that both thermoplastics and thermosetting plastics can be treated to give flexible and rigid cellular structures. The sizes of the cells (bubbles) or beads often control the density of the foam. Where the cells or beads are large, then the material will be of low density (because most of the volume is taken up with a gas). Manufacturers seek to obtain an even distribution of cells throughout the thickness of the sheet or moulding.

Because of their cellular structure and density, expanded plastics float on water, are good thermal and sound insulators and absorb impact energy well. They can therefore be found in buoyancy aids, in outdoor clothing, in cars as vibration-absorbent linings and crash padding, as well as in packaging. They have many structural applications where light weight and high strength are important.

The majority of these expanded plastics, with the exception of expanded polystyrene (EPS) are made by a continuous open-casting process in which the raw material is sprayed on to a walled conveyor belt and allowed to expand freely under heaters, like a large loaf of bread. When the material has cured it is sliced into sheets and blocks. Plastics prepared in this way show their cellular structure and can be either rigid or flexible.

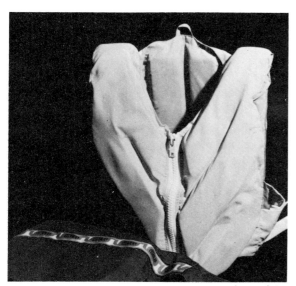

Flexible foam EVA (ethylene vinyl acetate) used to provide buoyancy to these canoeists' lifejackets.

Expanded injection-moulded or extruded plastics have a hard or tough natural skin of higher density that copies the mould surface.

Good-quality products made from rigid expanded plastics are often sprayed with special paints, or textured, to hide the 'frozen-in' flow lines created by the cells moving in the mould during processing.

Occasionally one can find articles made of expanded plastics that look remarkably like pieces of wood. Close examination shows that the wood grain appearance is caused by the stretching of the individual cells in the material. It is highly probable that in the future this 'moulded-in grain' will be used to represent wooden products.

The following pages show how expanded and foamed plastics can be used in schools. Some rigid slab (thick sheet) can be carved, while some flexible slab can be cut and used for upholstery. Two types of mouldable expanded thermoplastic sheet materials are available for strip bending, oven and vacuum forming. A further type of expanded polystyrene is available in bead form for expanding into mouldings. *Take note of the safety precautions outlined in each section.*

Expanded polyurethane

Rigid expanded polyurethane is available in sheets and blocks in a wide range of sizes, thickness, and densities. It is generally used as a building material for thermal insulation and sometimes on boats and in canoes for buoyancy.

In recent years many schools have used it as a quick-modelling medium. It is unpleasant to handle but can be carved easily and accurately with an old hacksaw blade and lightly filed to almost any form (Figure 8.2A).

Warning: The dust created by sawing and filing must be kept to a minimum, and it is wise to work over a wet cloth. Wear a face mask and vacuum-clean the area constantly, or work out of doors, away from other people.

Rigid polyurethane foam has been used in the entire structure of this desk with the exception of the front tubular leg and fixing bolts.

This polystyrene bowl has been injection moulded but during the process the hot plastic melt has had gas introduced into it creating a cellular structure.

Closed cell polystyrene has been used for this vacuum formed egg carton.

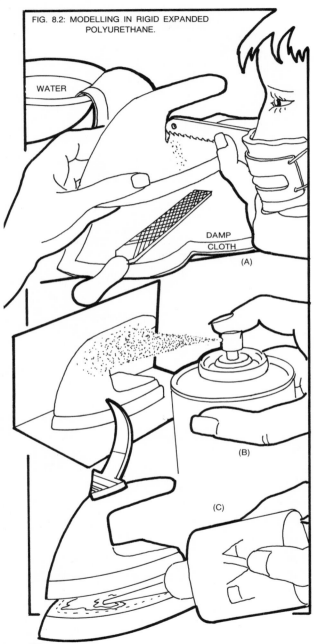

FIG. 8.2: MODELLING IN RIGID EXPANDED POLYURETHANE.

WATER

DAMP CLOTH

(A)

(B)

(C)

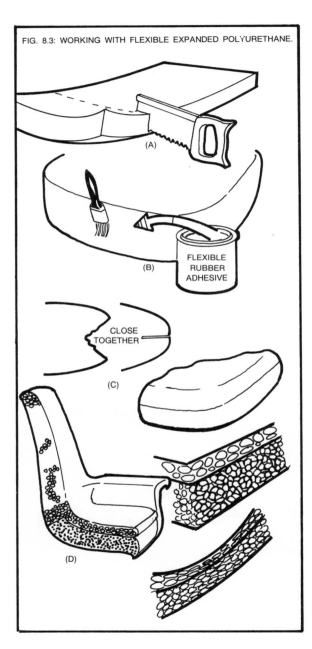

FIG. 8.3: WORKING WITH FLEXIBLE EXPANDED POLYURETHANE.

(A)

FLEXIBLE RUBBER ADHESIVE

(B)

CLOSE TOGETHER

(C)

(D)

Models are best made in sections, sprayed with car aerosol paints in a spray booth (Figure 8.2B) and then assembled and cemented together (Figure 8.2C). PVA wood glue works well but should not be taken to the very edge of a joint, because when dry it is harder than the foam and cannot be cleaned back easily.

Warning: never use a hot-wire cutter on these polyurethane materials. This will cause very dangerous fumes to be given off.

Polyurethane foam

Flexible polyurethane foams are available in different forms and at very different prices. Variations occur in quality, density, and composition.

When buying flexible polyurethane foam for upholstery applications it is important to insist on a *fire-retardant* type. Buy two different densities. Low-density foam is best for chair backrests and as a surface skin over a thick layer of high-density foam for seat squabs (Figure 8.3D). It gives a soft gentle sensation of comfort when one sits down but compresses easily. The high-density foam 'gives' a little but provides support for long periods. If these foams burn they give off very dangerous toxic fumes — fire-retardant versions must be used. They can be cut easily with a fine tenon saw (Figure 8.3A) and glued with a spirit-based rubber adhesive (Figure 8.3B). Figure 8.3C shows how radiused edges may be achieved by glueing, drying and folding edge-to-edge to give a professional finish.

Polyurethane plastics can be formulated by the industrial chemist to produce either thermoplastics or thermosetting plastics. Polyether and polyester expanded plastics are two types of polyurethane materials used for upholstery purposes and foam-backed fabrics.

Plastazote

This is a flexible, skinless closed-cell cross-linked polyethylene material. It is available in white, in sheets in several thicknesses from 3 mm to 25 mm, and in several primary colours and black in 6 mm thickness. It is commonly used for buoyancy in lifejackets and for swimming aids, and has many medical and packaging applications.

Although crosslinked, it is still a thermoplastic and can be heat press-formed, vacuum-formed and heat-welded (Figure 8.4). It is a good thermal insulator (and therefore a poor conductor of heat), making it important to heat it very slowly before forming. The material will mould at 100°C.

Plastazote can be locally heated and 'pinched' together to enable a hinge effect to be achieved.

Foamex or simocel

Foamex (closed-cell expanded rigid PVC) is a relatively new material. It is available in several thicknesses, in grey or white, can be formed on a strip heater or vacuum former, and can be worked with woodworking tools. It will float on water (density 0.7), will not rot, is tough and impact-resistant, and can be pinned provided the pins are not closer than 6 mm to the edge of the sheet. The surface skin of the material can be painted or screen-printed but is very thin. Because the skin is thin, vacuum-forming moulds must be designed with generous tapers and large radii over shallow depths of draw. It must be heated slowly before forming (ideally on both sides). Overheating causes the material to 'yellow' and blister. Foamex can be cemented or heat-welded. Figure 8.5A shows 'poor' vacuum forming mould shapes, and Figure 8.5B shows preferred ones.

Expanded polystyrene

Expanded bead polystyrene is common in packaging, as ceiling tiles and in slab, sheet and

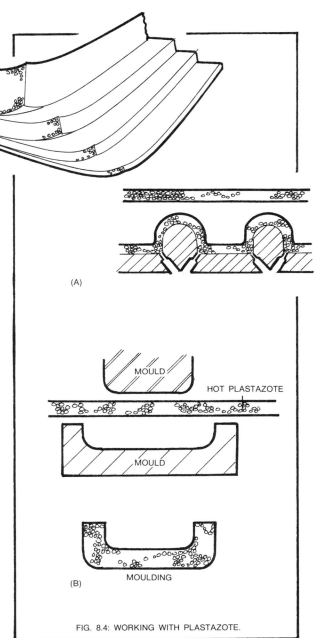

(A)

MOULD

HOT PLASTAZOTE

MOULD

MOULDING

(B)

FIG. 8.4: WORKING WITH PLASTAZOTE.

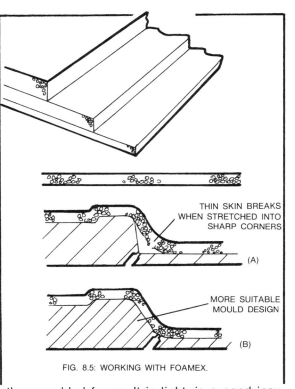

THIN SKIN BREAKS WHEN STRETCHED INTO SHARP CORNERS

(A)

MORE SUITABLE MOULD DESIGN

(B)

FIG. 8.5: WORKING WITH FOAMEX.

other moulded forms. It is light, is a good insulator and is easily recognized by the pearly white beads that make up its structure. The standard tool for cutting is a heated nichrome wire, which gently melts its way through the material.

Warning: Hot-wire cutters should be used only at black heat to avoid giving off fumes and smoke.

The basic steps for making a block of expanded bead polystyrene are shown in Figure 8.6A–E. When first obtained, the beads are minute — the size of a pin head — but contain a gas that expands at the softening temperature of polystyrene (90°C). The beads are heated loose in a

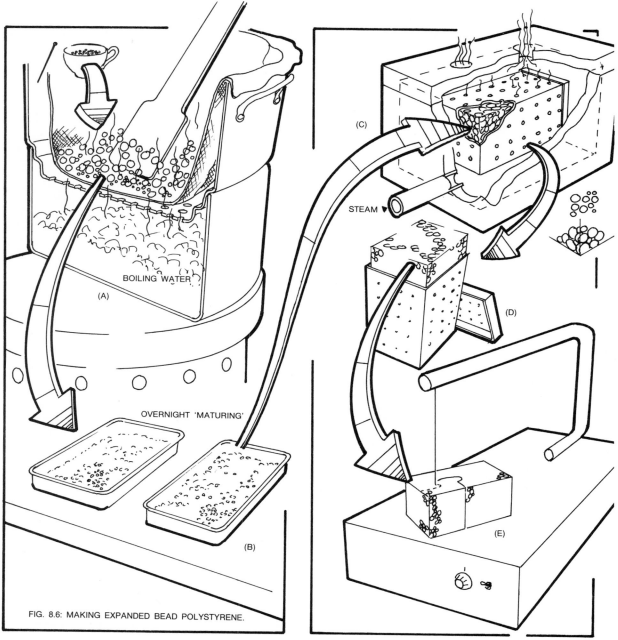

STEAM ▼

BOILING WATER

(A)

OVERNIGHT 'MATURING'

(B)

(C)

(D)

(E)

FIG. 8.6: MAKING EXPANDED BEAD POLYSTYRENE.

steam chamber (Figure 8.6A) — that is, in a fine sieve over a boiling-water bath — occasionally being agitated by a wooden paddle. The heat softens the polystyrene and the gas slowly expands within the material, causing the bead to get larger and larger — up to 50 times its original size.

The beads are left to cool and 'mature' for 12 hours (Figure 8.6B). They are then packed into a finely perforated metal mould, which is filled completely and firmly capped (Figure 8.6C). The mould is placed into a steam chest and steamed. For experimental purposes use a pressure cooker on its lowest pressure setting. The amount of steam depends on the mould volume. Again the beads soften with heat and the gas expands further, but this time, because the beads are already tightly packed, they press firmly against one another, expand into the spaces and lightly weld together. When cool, a solid block of tightly packed beads can be removed (Figure 8.6D). This block can now be shaped using a nichrome hot-wire cutter (Figure 8.6E). When you use this tool, it is important to remember that the wire should only be hot enough to melt the polystyrene. The polystyrene *must not be permitted to burn or smoke*. If it does, reduce the power by means of the variable temperature control (if fitted), or reduce the voltage. The usual working voltage is 12 volts.

A different expanded polystyrene is also available as a closed-cell semi-rigid sheet material, suitable for vacuum forming and also used for packaging — for example, in some egg boxes. Expanded plastics materials are becoming increasingly important and many developments are likely in the next few years.

Twin skin corrugated plastic sheet

Several versions of these materials are available in sheets 3 mm thick × 1 m × 2 m in a wide range of bright colours. The material has two thin outer skins interconnected with fine ribs spaced 3 mm apart running parallel along its length. The material is usually polypropylene and because the sheet is hollow it is very light in weight, relatively strong and unaffected by water. It appears very similar to corrugated cardboard but is stronger, more rigid, more impact and abrasion resistant.

Although it is thermoplastic the material does not heat form or weld easily but requires cutting, folding and clipping with plastic or metal fasteners. To make a box container from the material it is best to make a full size template from thin card, fold it to shape and work out the fixing points. Then use the template to mark out and cut the plastic sheet. The material will fold easily along the line of the ribbing but is more difficult across the ribbing. A leatherworker's rotary wheel stitch marker will make a line of fine holes slightly weakening the material to make it fold more easily. When folding the sheet it is sensible to use a steel rule above and below the sheet as a guide and support to achieve a clean bend. Use a hairdryer or strip heater to *slightly* warm the material in the bent position, allow it to cool and it will hold its shape.

Large headed fabric and leather working fittings and rivets can be used as fasteners but some injection moulded fittings are available. The photograph shows a briefcase made from this material with folded tabs and rivet fasteners. This has been in daily use for two years and shows little sign of damage.

PROJECTS AND ACTIVITIES

(1) Why are foam materials often used for packaging? Devise tests to examine the best foam to use for packaging an egg or a wine glass.

Then design and make a package using the minimum foam material to pack an egg. Drop the package and egg from table height (76 cm) to the floor. Can your design be improved upon?

(2) Design a buoyancy aid for use in a swimming pool. State (1) what type of expanded material you would use, (2) how you would form it and (3) why it is particularly suitable for use by young people in a swimming pool.

(3) Devise an experiment that will enable you to work out the weight of polystyrene bead foam necessary to keep an adult person afloat in sea water. Would you need more or less foam for the same purpose in a river?

A two year old briefcase made from corrugated polypropylene sheet, cut folded and held together with specially made injection moulded rivets and fittings. The case has withstood the knocks and scratching of daily usage very well.

(4) Devise a test to examine the heat conductivity of a foam.

(5) Design a rig to examine the transmission of sound through a foam.

(6) Suggest ways in which foams can be joined to themselves and to other materials. State what you would use and why. Try out your ideas and compare your results.

(7) Discover in your home as many items using plastics foams as you can. List them and describe their purposes.

(8) Can you find examples of a thermoplastic foam, a thermosetting foam, a flexible foam, a rigid foam, a bead foam, an open cell foam and a closed cell expanded material? In what way are their properties used to the benefit of the product?

(9) Are there any dangers from the use of foam in furniture? How could you *safely* find out?

(10) Felt pens, pencils and other graphic aids can be difficult to locate at the bottom of a briefcase or holdall. Examine the items you use most in preparing artwork and design engineering drawings and then design a series of interlocking trays that can be vacuum formed to house your equipment. Design a method for holding these trays in an outer case made from corrugated sheet, or Corriflute polypropylene fluted sheet.

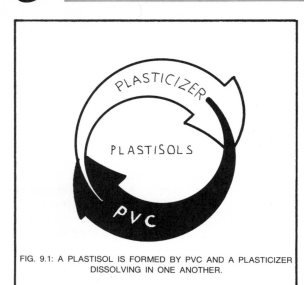

FIG. 9.1: A PLASTISOL IS FORMED BY PVC AND A PLASTICIZER DISSOLVING IN ONE ANOTHER.

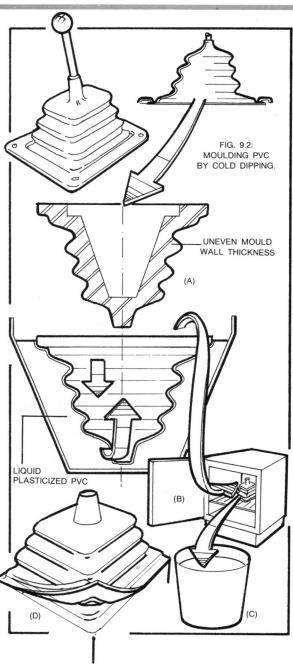

FIG. 9.2: MOULDING PVC BY COLD DIPPING.

UNEVEN MOULD WALL THICKNESS

(A)

LIQUID PLASTICIZED PVC

(B)

(C)

(D)

PVC (polyvinyl chloride) in its pure chemical state is a hard, colourless, heat-sensitive, almost unmanageable material, quite unsuitable for general conversion into products by the usual thermoplastics processes. To overcome these problems chemists have formulated materials that can be added to PVC during manufacture to enable products to be made ranging from injection- and extrusion-moulded rigid products to soft leather-look flexible fabrics, flexible foams, and tough, flexible hard-wearing shoe soles. These additive materials are formulated with PVC in different proportions to control both processing and end-product properties. The main types are *plasticizers*, *stabilizers*, *fillers*, *pigments* and *lubricants*.

Stabilisers

Some of these chemicals can prevent PVC from burning when it is heated to moulding temperature, and can be used to prevent ultraviolet light (in sunlight) from making the PVC become brittle.

Fillers

These are usually added to reduce costs or to improve a particular technical quality, for example surface hardness or electrical insulation properties. Typical fillers are china clay, talc, calcium carbonate and silica. *Pigments* are a further range of chemicals added to give colour.

Lubricants

Lubricants are a range of chemicals that are mainly added to a mix to prevent it from sticking to hot metal surfaces during processing.

A product manufacturer can buy a wide variety of PVC materials pre-prepared, containing these ingredients in specified proportions or have bulk quantities prepared to his or her specification.

PLASTISOLS

An important group of materials that can be bought as liquids and pastes are called *plastisols*. The remainder of this section deals with their simple but unique processing methods. A plastisol is a mixture of PVC and a plasticizer, the two materials having been dissolved into one another (Figure 9.1). The plasticizer will be a solvent or a mineral oil which will be absorbed into the PVC particles which become swollen in time. When the resulting plastisol (liquid or paste) is coated onto a metal surface and heat-treated, it produces a tough, chemically resistant, hardwearing skin. This skin (of plasticized PVC) may be removed as a finished product, or it may remain as a permanent coating. The heat treatment does not cause a chemical change of material composition but only modifies the plastisol, causing it to change its physical state from a liquid to a flexible solid.

Plastisols are used to produce coatings, mouldings, forms, and flexible mould-making materials. The following methods are shown: cold dipping and hot dipping, spreading, low-pressure casting, slush moulding, rotational casting, for producing coatings and mouldings; hot-melt casting to produce flexible moulds and for Vinagel modelling.

These materials are not suitable for use in schools, except for the hot-melt casting of Vinamold and modelling of Vinagel.

Cold dipping

This process is used to coat moulds or metal articles with uneven wall thicknesses. The mould is a positive shape and is lowered very *slowly* into a container of PVC plastisol until completely covered (Figure 9.2A), allowed to rest or 'dwell' for a short period of a minute or two, then raised very slowly and placed in an oven (Figure 9.2B). Dipping and removal are carried out slowly to prevent air being trapped and to stop runs of PVC developing. When placed in the oven, the mould is turned upside down to prevent drips developing. Heat curing in the oven takes place at 150–180°C for 10 minutes. The mould is then dipped into a water-bath at 60°C (Figure 9.2C). When cooled to this temperature, the PVC will stretch to lift off complicated undercuts and return to its original moulded shape. The complicated mould shown (Figure 9.2D) gives some idea of the extreme forms possible with this process. *Double dipping* is possible if carried out during the oven curing period. Material thicknesses of up to 3 mm are possible.

Hot dipping

This process is similar to cold dipping, but is used when the metal mould has a constant wall thickness. In this case the mould is heated to 110–130°C in an oven (Figure 9.3A) before being slowly lowered into a container of PVC plastisol (Figure 9.3B). Again it is removed very slowly, after a short dwell time during which material adheres to the surface. Then it is turned upside-down and returned to the oven (Figure 9.3C) to cure at 150–180°C. The length of dwell time in the plastisol will govern the coating thickness — the shorter the time, the thinner the coating. Other factors that affect the coating thickness are: (1) the fluidity of the plastisol; (2) the mould temperature and its ability to hold heat; and (3) the mould shape. When cured, the mould and its coating are lowered into a water bath at 60°C (Figure 9.3D); then the moulding is removed.

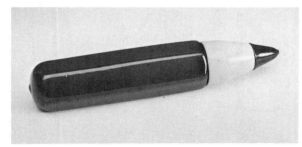

Plastic pencil case manufactured by hot dipping in PVC plastisol. The cap of the pencil case is produced in two colours by double dipping in two different colour solutions of PVC plastisol.

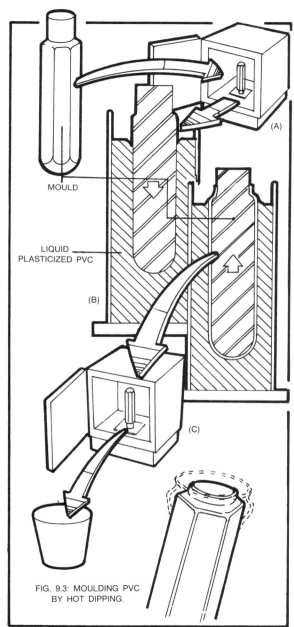

FIG. 9.3: MOULDING PVC BY HOT DIPPING.

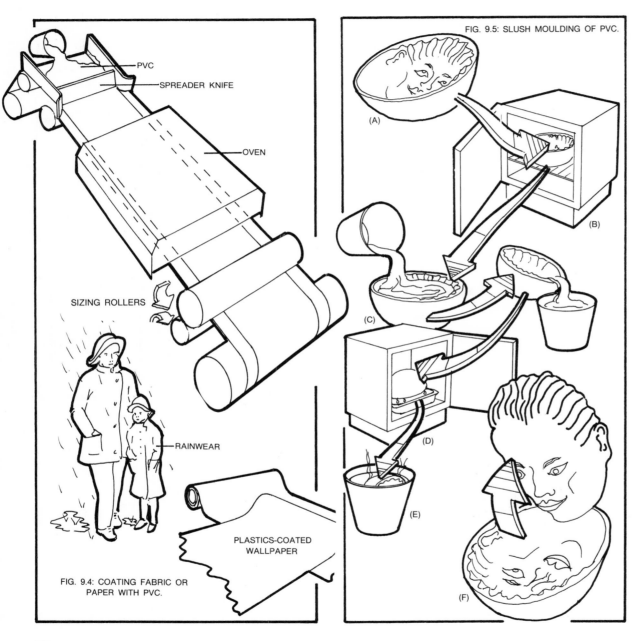

FIG. 9.5: SLUSH MOULDING OF PVC.

(A)

(B)

(C)

(D)

(E)

(F)

PVC

SPREADER KNIFE

OVEN

SIZING ROLLERS

RAINWEAR

PLASTICS-COATED
WALLPAPER

FIG. 9.4: COATING FABRIC OR
PAPER WITH PVC.

At 60°C the material is capable of being stretched over undercuts and will return to the original mould profile.

Spreading

Occasionally it is necessary to apply a PVC coating to paper or fabric. Plastic-coated wallpaper, upholstery leather-look fabric and rainwear fabrics are produced by a sophisticated version of this method. In principle the paper or fabric is stretched between two main rollers, one carrying the supply of material, the other taking up the coated, cured product (Figure 9.4). As the paper unwinds from the supply roll it collects *plastisol paste*, which is spread in an even thickness by a doctor blade. The paper, now carrying an even coating of paste, passes through a curing oven at 165–200°C, which causes the paste to gel into a dry flexible coating. At this stage a textured pattern can be impressed into the coating by a pair of embossed nip rollers. These are made of steel and run at room temperature. Fully cooled and cured, the coated paper is finally collected and rolled up in suitable lengths, ready for use.

Slush moulding

This is an open-mould casting process. The mould is a hollow bowl made of metal, similar to a jelly mould (Figure 9.5A). It is heated to 100°C (Figure 9.5B); then the PVC plastisol is slowly poured in and allowed to stand for a 'few minutes (Figure 9.5C). The PVC is then poured out, leaving a coating over the inside surface. The mould is returned to the oven (Figure 9.5D) for curing at 150–165°C. Finally it is cooled in a water bath (Figure 9.5E) and the moulded article is stripped out (Figure 9.5F).

This process is rarely used commercially these days because it has been superseded by *rotational casting*.

Rotational casting

In recent years this has become an important commercial process for the production of hollow thermoplastics products. It is a slow process, but products can be made rigid or flexible, ranging in size from dolls (limbs, bodies and heads) to chairs and canoes. It is used to make products from either PVC plastisols or polyethene powders.

A special oven is required, incorporating a mechanism that is used to revolve the mould slowly through *two planes* (*in two directions*) *at the same time* (Figure 9.6). The mould is made of metal; it is hollow and completely encloses the cavity. Before the process starts, the mould is loaded with a measured amount of PVC plastisol or polyethylene powder (Figure 9.6A) and is then fitted to the rotator mechanism. The metal mould transmits the oven heat as it rotates. The plastisol is cured during this procedure. If polyethylene powder is used, it melts and flows throughout the mould cavity, completely coating the interior surface with an even thickness of material. The speed of rotation in the two planes governs the distribution of material thickness. After about ten minutes the mould is quickly immersed in cold water (Figure 9.6B) and the moulding removed. A flexible moulding that has deep undercuts, such as the doll's head shown, needs to be punctured (Figure 9.6C), so that the air is allowed to escape. It can then be collapsed before removal. Rigid mouldings, or mouldings that cannot be punctured, such as footballs, have to be made in split moulds, which are opened up for the removal of the product. The mould split line is often visible on articles made in this way. Sometimes two articles are made in a combined cavity mould; for example, two road cones can be rotationally cast base to base at the same time. The double-ended moulding is then sawn in half to produce two products.

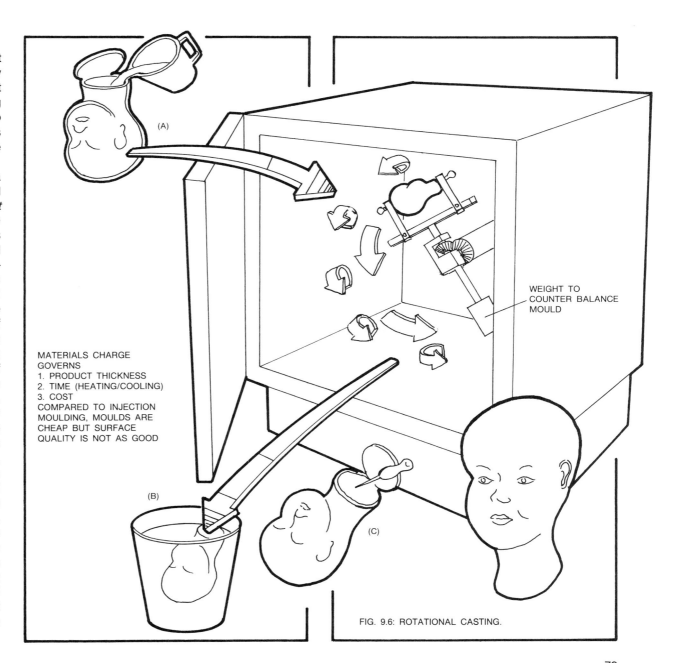

(A)

WEIGHT TO COUNTER BALANCE MOULD

MATERIALS CHARGE GOVERNS
1. PRODUCT THICKNESS
2. TIME (HEATING/COOLING)
3. COST
COMPARED TO INJECTION MOULDING, MOULDS ARE CHEAP BUT SURFACE QUALITY IS NOT AS GOOD

(B)

(C)

FIG. 9.6: ROTATIONAL CASTING.

FIG. 9.7: WORKING WITH VINAMOLD.

(A)

(B) AIR GAP BETWEEN PAN AND HEATER

ORIGINAL

METAL TUBE

(C)

(D)

POLYESTER RESIN OR PLASTER MIX

(E)

(F)

COPY

ORIGINAL

80

Vinamold

This is a flexible, thermoplastic plasticized PVC material, available in several grades for mould-making. Each grade has a different degree of flexibility; the range most suitable for use in schools is denoted by colours, red and yellow. Red is soft and flexible and looks like raw meat, while yellow is harder and only just flexible. The materials are suitable only for mould-making on a small scale. The choice of grade to use is based on (1) mould size, (2) mould shape and complexity, and (3) the medium to be cast into the finished mould (its weight and the probability of distortion).

The Vinamold should be cut into cubes 10 mm in size (Figure 9.7A), heated in a purpose-made boiler or pan, and *stirred all the time*. Heating should be done in a fume cupboard or under an extractor hood, and the pan should not touch the hot-plate directly but have an air space between to prevent burning (Figure 9.7B). Melt temperature for both materials is 150–170°C, but they should be poured at 150°C.

Warning: Heat-proof leather gloves, apron and safety goggles must be worn for this operation.

When molten, the Vinamold should be poured to one side of the object to be moulded, to prevent air being trapped (Figure 9.7C). The object should be sealed if it is made of a porous substance such as plaster. The Vinamold holds its heat for a long time, so it should be left to cool overnight and stripped only when cold.

When cold, the original object is removed and the new negative mould (Figure 9.7D) is turned upside-down, ready for casting a copy of the original. When polyester resin is cast into the mould it is sensible to let the resin flow down a rod (Figure 9.7E) to let air flow out. The casting is left for 24 hours to cure before it is removed.

The Vinamold can then be chopped up for re-melting to make another mould (Figure 9.7F).

The material is self-releasing, so it does not require release agents before polyester resin casting. Plaster and concrete can also be cast into Vinamold. Any burnt or contaminated material should be discarded.

Particularly large castings can be made by first covering the original with a 20 mm-thick clay coat. Then plaster is cast over the clay in another 20 mm-thick coat. When hard, the plaster and clay are removed and the plaster and the original model are cleaned. When dry the two are assembled upside-down and Vinamold is poured into the space made by the clay.

To make the final cast the Vinamold skin is supported inside the plaster skin and the mould cavity is filled with resin or the casting medium.

Vinagel

Several grades of this modelling material are available, from which either flexible or rigid products can be made. It is another form of plasticized PVC, in appearance white and crumbly, closely resembling pastry dough. It can be handled and worked until soft and pliable (Figure 9.8A), modelled to shape with box-wood modelling tools (Figure 9.8B) and finally, to make the model permanent, it can be fired in an oven at low temperature (150–165°C) until cured (Figure 9.8C). The firing drives off the excess plasticizer, leaving a flexible or rigid solid model, depending on the grade. The curing time is dependent on the bulk of material used — the greater the quantity, the longer the time — ranging from half an hour to several hours. Cured material becomes translucent or milky in appearance. The finished object can be painted using PVC, acrylic or oil paints.

Small quantities of turpentine can be added to thin material that has dried out, or household starch can be added to thicken it.

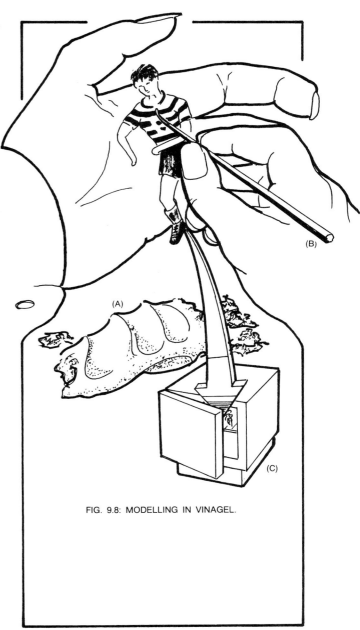

FIG. 9.8: MODELLING IN VINAGEL.

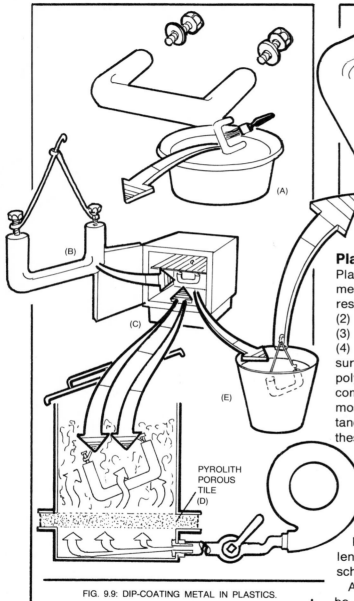

PYROLITH
POROUS
TILE
(D)

FIG. 9.9: DIP-COATING METAL IN PLASTICS.

Plastics powder coatings

Plastics are used to provide surface coatings on metals for the following reasons: (1) to improve resistance to weather and chemical corrosion; (2) to provide electrical and thermal insulation; (3) to create low-friction non-stick surfaces; and (4) to provide attractive, colourful hard-wearing surfaces. The main thermoplastics used are polyethylene, nylon, PTFE and some PVC compounds. Epoxy powders are the main thermosets used to provide good chemical resistance and tough durable finishes. Commercially these materials are applied either by electrostatic powder spraying or by fluidized-bed powder coating. Both systems achieve the same end-result of providing a plastic skin over the exposed area of metal.

Here is how to plastics dip-coat a metal article, using either powdered polyethylene or nylon and a fluidized bed. Polyethylene and nylon powders are the most suitable for school use.

All oil, grease, old paint and rust scale must be removed to ensure good adhesion of the coating. The articles should be scrubbed in hot soapy water (Figure 9.9A), if a degreasing agent is not available, and dried thoroughly.

Fix a suspension wire (Figure 9.9B) to a suitable point, preferably one that will not be coated or that will be hidden when the component is fitted to the rest of the product. Mask all bolt holes, or fit suitable bolts inside the holes to prevent coating the thread. The bolt may be an ideal suspension point for attaching the wire. Suspend the object in an oven (Figure 9.9C) with a drip tray below it to catch excess drips that may occur from the later coating. When thoroughly heated, the object should be removed from the oven and slowly lowered into the fluidized powder (Figure 9.9D). Some particles of powder will melt on and stick to the object, forming a fine layer over it. It is unlikely that the object will be covered in a complete film, so reheating is necessary to help the powder flow. Re-dip the object in the fluidized tank to increase the coating thickness. The powder will stick more readily to the hot plastics film already covering the object. Reheat the coated object in the oven to fuse the plastic fully into a continuous layer. Avoid leaving the object hanging in the oven for longer than necessary because the hot plastic will run and drip off the lowest points leaving a thicker coating lower down.

When an even coat has been achieved, remove the hot plastics-coated object from the oven and cool it in a water bath (Figure 9.9E).

If the hot plastics material is allowed to touch the oven or dip-tank sides, the coating will be spoilt. Great care is necessary when handling the object, and gloves must be worn to avoid serious burns.

Only metals can be plastics-coated by this method. Do not coat springs or hardened steel cutting edges or soft soldered wires.

PROJECTS AND ACTIVITIES

(1) Find ten examples of plasticized PVC in your home. When looking for them, what particular qualities do you expect to find to help identify them?

(2) What are the problems related to the design, construction and materials of articles for plastics coating using fluidized-bed powder methods? Carry out a series of experiments to show how these can be overcome.

(3) Many machine tools and stored steel items of military equipment are coated in PVC plastisol. Can you suggest any reasons for this?

(4) The handles of electrical pliers and the shafts of electrical screwdrivers are often coated in PVC plastisol. Is this just to improve their appearance or does it serve some useful functions? What are they?

(5) Obtain a pair of domestic PVC gloves, the type used for washing-up and doing dirty jobs around the home. Describe how you think they have been made, paying particular attention to the surface finish.

(6) Obtain two articles, one which you know to have been made by rotational casting and the other by extrusion-blow moulding. Compare the two closely, describe what you see and sketch any significant details that would help a less knowledgeable person identify which product had been made by which process.

(7) Is it only PVC plastisol that can be used for rotational casting? If not suggest another material and find examples of both. Is there a noticeable difference in the rigidity of the mouldings?

(8) PVC materials can range from being very hard to very soft. Collect a wide variety of small samples and make up a chart for your classroom to show the different grades of PVC. Show how the designer takes advantage of the different degrees of hardness/softness for making better products.

(9) Devise an experiment to measure the difference between hardness and softness in different samples of PVC of the same thickness and size.

(10) Many different materials are added to PVC materials to change their properties. Why is this necessary and do these changes affect processing?

(11) Obtain a child's doll that is no longer wanted made from flexible PVC. Take it to pieces and if possible carefully cut it up so that you can see the thickness of the material. Is the thickness constant, and are there any 'tell tale' lines on the moulding surfaces which might suggest the use of split moulds? Compare these doll mouldings with sections cut out of a washing-up liquid squeezy bottle. Have the doll's parts been made by the same process as the squeezy bottle, how do the sections compare, and what are the essential moulding differences between the two types of product?

(12) Can you find any examples of paper, fabric or other materials being used as reinforcement or supporting material for PVC coatings? Examine each sample and suggest reasons why the designer has used PVC as the surface coating material.

(13) PVC cold and hot dip coating are common processes used for the manufacture of a variety of different products. Find product examples of each and make drawings to show the probable mould design for each process.

(14) Plasticised PVC is used for several parts in a car interior. How many examples can you find and what processes were used to make each part?

(15) Using a small drill brace, can you make a device to simulate the motion of a rotational casting machine.

(16) Rotational casting is often used to make large hollow toys. Find an example and examine it closely. Draw the product and show how the mould might have been designed.

(17) Describe the process of slush moulding and compare it to slip casting in ceramics. In what way are the processes similar and how do they differ? Write a brief illustrated description of each process.

(18) What common products are made from materials that have originally been calendered? Obtain some small samples and prepare them so that they show the surface layer and the base material.

(19) Many modern electronic fashion watches have casings and straps made from plasticised PVC. Remove the electronic mechanism from an old one that no longer works and using this as a basis for a redesign fashion styling exercise produce some visually exciting new design ideas that could be made by injection moulding of flexible plasticised PVC. Make an accurate full size model from Plasticine, Vinamold or Vinagel. Then design a point of sale display package.

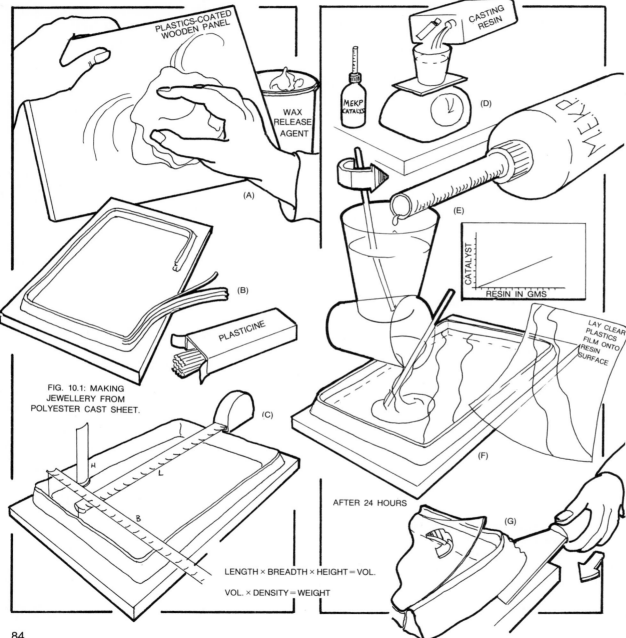

FIG. 10.1: MAKING
JEWELLERY FROM
POLYESTER CAST SHEET.

LENGTH × BREADTH × HEIGHT = VOL.

VOL. × DENSITY = WEIGHT

JEWELLERY FROM CAST SHEET

Polyester jewellery can be made by casting sheet, in this way. Prepare a melamine-surfaced board by applying a release agent (wax) (Figure 10.1A). Carefully build a wall of Plasticine around the edge. Work the Plasticine neatly and thoroughly on to the melamine board to form a watertight reservoir (Figure 10.1B). This mould can be used to produce:

(1) clear cast sheet;
(2) clear cast sheet with swirls of translucent pigment;
(3) translucent coloured sheet;
(4) opaque coloured sheet.

To produce a clear or translucent coloured sheet, measure the length, breadth and depth of the sheet in centimetres, multiply these dimensions to find out the volume of resin required (Figure 10.1C) in cubic centimetres. Using a measuring cylinder marked in cm³ (or millilitres, ml) measure out the exact number of cm³ of water into a polyethylene mixing cup. Stand the cup on a horizontal surface and mark the water level. Pour away the water and thoroughly dry the cup. When dry, pour in casting resin, up to the mark. Weigh the resin (Figure 10.1D), then add pigment — drop by drop to make a translucent coloured sheet, or up to 7 per cent by weight, if opaque coloured sheet is required; alternatively, leave clear. Then add the catalyst (Figure 10.1E) according to the manufacturer's instructions.

Warning: Methyl ethyl ketone peroxide (MEKP catalyst) is highly dangerous and flammable — it must be dispensed from a properly calibrated safety dispenser. Always wear eye protection — if splashes enter the eye, get medical attention immediately. Wear barrier cream and disposable polyethylene gloves.

To achieve a cast sheet with a second colour cast in, use the following method. Fill the mixing cup to the mark, weigh it, and then catalyse all the resin with the correct proportion of catalyst. Take a second mixing cup and pour in one-eighth of the catalysed resin. Add some pigment to one or both resin mixes.

Carefully pour the larger resin mix onto the melamine board mould and check that the resin spreads to an even depth over the complete area (Figure 10.1F). Then carefully pour the second resin mix onto the first in patches, giving separate areas of the second colour.

Finally, place a clean piece of stiff card over the mould to prevent dust falling on the resin and then leave for several hours to cure. When you return the resin may be solid but slightly sticky on the surface. Warm it to 50°C in an oven or on a radiator for 15 minutes, then allow it to cool. The sticky surface should now be cured. Remember that the resin is designed to be sticky on the surface to provide a 'chemical key' should you wish to cast another resin layer.)

To remove the cured cast resin sheet from the mould, gently ease the tip of a scraper between the resin and the melamine-faced board (Figure 10.1G). Take very great care because the resin is very brittle. As the resin lifts, slide a thin piece of rigid plastic sheet between the casting and the mould, taking care not to bend the casting. If the mould was properly prepared the two should now separate easily.

Select and cut out the areas of greatest interest and suitability for a pendant. Carefully file to shape, rub down with dampened wet-and-dry paper, and finally polish.

A second method is shown in Figure 10.2 for producing a cast sheet with two good moulded surfaces. By casting and curing in stages and tilting the mould, clever patterns can be achieved.

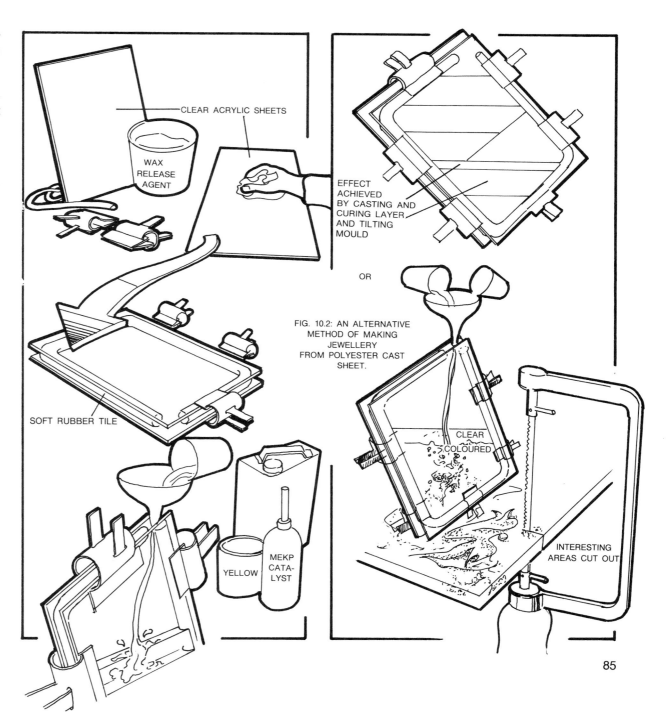

CLEAR ACRYLIC SHEETS

WAX RELEASE AGENT

EFFECT ACHIEVED BY CASTING AND CURING LAYER AND TILTING MOULD

OR

FIG. 10.2: AN ALTERNATIVE METHOD OF MAKING JEWELLERY FROM POLYESTER CAST SHEET.

SOFT RUBBER TILE

YELLOW

MEKP CATALYST

CLEAR

COLOURED

INTERESTING AREAS CUT OUT

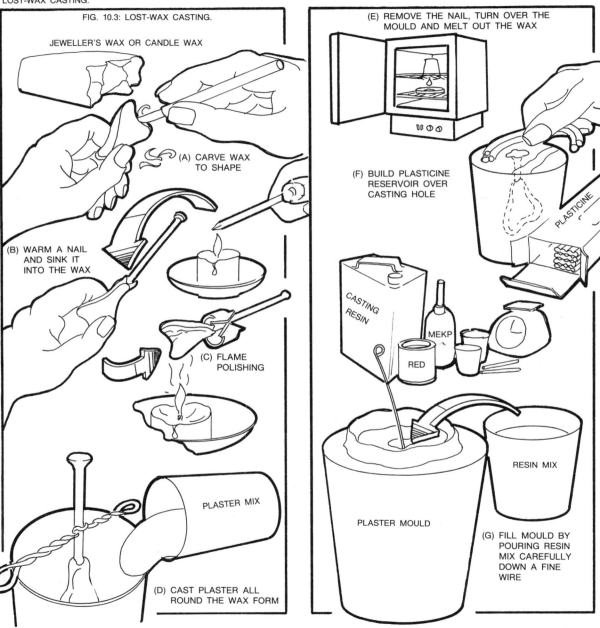

FIG. 10.3: LOST-WAX CASTING.

JEWELLER'S WAX OR CANDLE WAX

(A) CARVE WAX TO SHAPE

(B) WARM A NAIL AND SINK IT INTO THE WAX

(C) FLAME POLISHING

PLASTER MIX

(D) CAST PLASTER ALL ROUND THE WAX FORM

(E) REMOVE THE NAIL, TURN OVER THE MOULD AND MELT OUT THE WAX

(F) BUILD PLASTICINE RESERVOIR OVER CASTING HOLE

PLASTICINE

CASTING RESIN

MEKP

RED

RESIN MIX

PLASTER MOULD

(G) FILL MOULD BY POURING RESIN MIX CAREFULLY DOWN A FINE WIRE

LOST-WAX CASTING

Small jewellery forms can be modelled in jeweller's wax, or even candle wax, using miniature scrapers, dentist's tools, or shaped pieces of wood or plastics sheet. (See Figure 10.3 A–F for procedures for making a wax model for a pendant, fob or brooch.)

To melt out the wax, turn the plaster cast upside-down over a tray or tin lid in an oven (Figure 10.3E) and heat it to 80°C for at least two hours. This leaves a hollow mould cavity and continues to thoroughly dry the plaster cast, well after the time when all the wax has melted and drained away.

Taking great care to not allow any bits to fall into the cavity, build a watertight 'well' of plasticine (Figure 10.3F) around the pouring sprue hole. This will provide a resin reservoir that will top-up the resin in the cavity as air escapes and the resin shrinks. Mix and pigment (colour) the resin as described on page 84, and then very carefully pour the resin *slowly* down a fine wire so that it travels down one side of the sprue passage (Figure 10.3G). A slow trickle entering the mould cavity allows much of the air to escape up the other side of the sprue hole. Top up the reservoir and check that it does not leak. To make sure that all the air has escaped, either tap the plaster gently to dislodge any remaining pockets, or, for very complex shapes, put the plaster mould in a vacuum box and lower the pressure for a few moments. Then leave to cure for several hours. If the position of the wax form was marked on the plaster following casting, then the plaster can be chipped and scraped away easily. As the plaster is cut away, wax-impregnated plaster soon becomes noticeable as one gets closer to the casting. This waxy plaster breaks away easily, leaving the original form. Clean, rub down and polish as before, using dampened wet-and-dry paper and cutting paste.

EMBEDDING OR ENCAPSULATION

This is usually a small-scale casting operation based on the method described on page 84 for casting a clear sheet of polyester resin.

In this case, several layers are cast to build up the thickness slowly. This is to prevent too much heat (exotherm) being developed by the chemical reaction, and to provide material for the object to stand on. The surrounding walls will therefore need to be higher; or a plain container can be used as a mould (Figure 10.4A).

Decide on whether the object should be visible from all sides, or whether it should sit on a coloured base. A shell or fossil resting on sand can be more interesting than when left 'floating' in the middle of a cast block. On the other hand, an insect specimen cast into a domed mould will be magnified, making details easy to see.

Apply wax to all mould surfaces and then mix and pour resin layer 1 (Figure 10.4B). Put a dust cover over and leave to cure (Figure 10.4C). When solid but still sticky, place the object on the surface of layer 1 (Figure 10.4D). If it is likely to float on the resin, then mix only sufficient resin for layer 2, to allow the specimen to just stand in the set position. A relatively heavy thin specimen (a coin, for example) may just be covered by layer 2. Allow layer 2 to cure, again protected by a dust cover.

With thin specimens, layer 3 may be the top layer, and should be sufficiently thick for cutting back and polishing. For objects that float, or are particularly thick, then layer 3 may be only a middle layer, and will need to be followed by other mixes. Each mix must be fully cured before the next layer is cast. It should be remembered that the temperature rises, owing to the chemical reaction *after the resin has gelled*, so gently feel the outside of the mould from

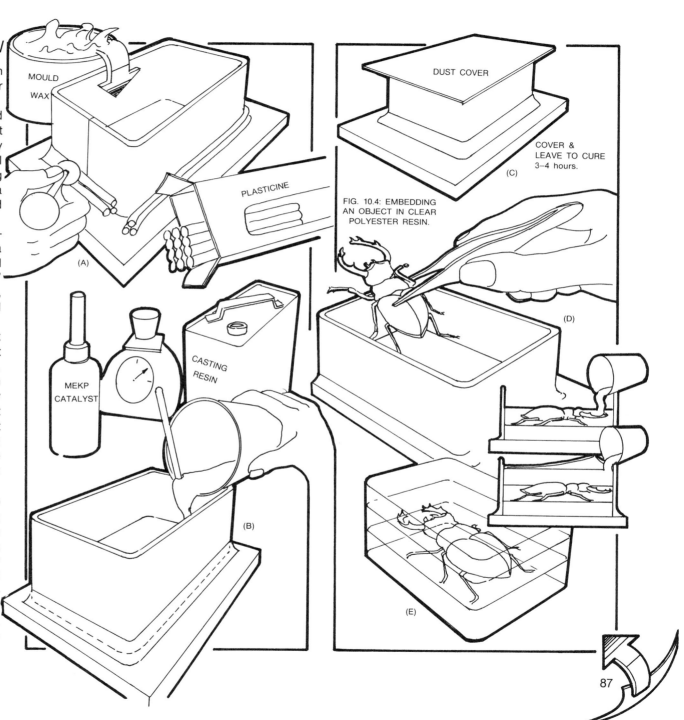

MOULD

WAX

PLASTICINE

(A)

MEKP CATALYST

CASTING RESIN

(B)

DUST COVER

COVER & LEAVE TO CURE 3–4 hours.

(C)

FIG. 10.4: EMBEDDING AN OBJECT IN CLEAR POLYESTER RESIN.

(D)

(E)

87

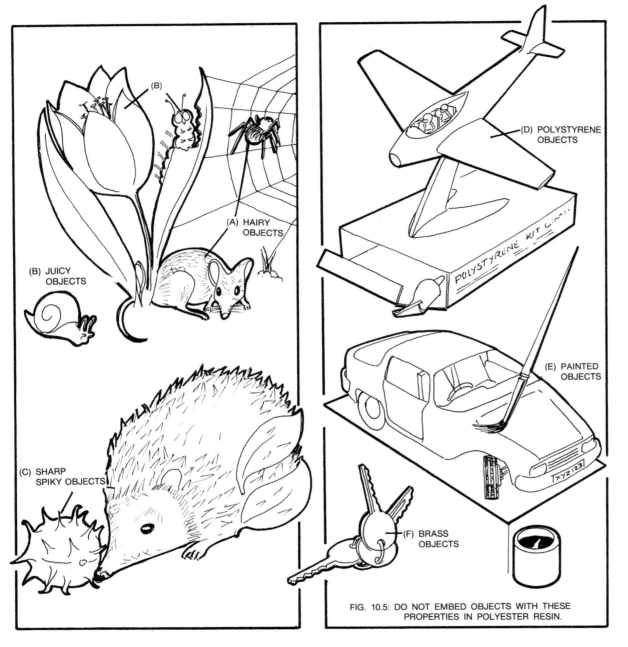

(A) HAIRY OBJECTS

(B) JUICY OBJECTS

(C) SHARP SPIKY OBJECTS

(D) POLYSTYRENE OBJECTS

(E) PAINTED OBJECTS

(F) BRASS OBJECTS

FIG. 10.5: DO NOT EMBED OBJECTS WITH THESE PROPERTIES IN POLYESTER RESIN.

time to time, and pour the next layer only after the previous one has cooled to room temperature. The finished casting should look like Figure 10.4E.

It is possible that the different layers will be noticeable from the sides. To avoid this it is essential that all casting and curing takes place in a dust-free atmosphere and it is helpful to always mix identical resin/catalyst quantities. Never pour the resin directly over the specimen, otherwise air will be trapped. Always pour down a stick to one side of the mould so that the resin level rises around the object.

The object to be embedded can be an insect, a flower head, a fossil, pieces of cast coloured resin, metal swarf, tin foil or fabric, the only precaution being that the object must be dry and unaffected by contact with the resin. Avoid hairy insects, flower heads that contain juices that cannot be dried out, and spiky objects. Also avoid polystyrene models, small painted objects or brass objects, as these will affect or melt in the resin. (Figure 10.5.)

There are some products on the market for preserving botanical specimen colours. Specimens should be dipped and then dried in a warm air draught before encapsulating in resin. A light mist of aerosol hair-lacquer or artists' spray fixative often works well, but these are not recognized methods and should be experimented with before use on important specimens. The light mist aerosol spray is particularly effective on the dusty wings of butterflies which are not fully wetted by the resin.

The hairs on many plants and insects trap air, preventing the resin penetrating these areas. These finished encapsulations have an attractive silvery appearance but are useless for showing the object. An artist's paint brush dampened with a drop of acetone and brushed over the insect can solve this problem.

LARGE CASTINGS

A large casting would be one requiring 0.5 kg or more of polyester resin. This may not appear to be very large; however this quantity of resin and the heat that will be generated by it require very careful control if satisfactory results are to be obtained.

Resins cure by chemical reaction. Most resins are supplied to suit the requirements of small-scale castings and have the correct proportions of accelerator already added, so that they need only pigment and catalyst to produce colour and solidity. When larger quantities of resin are needed, the proportion of accelerator and catalyst are reduced. The greater the volume of resin used *the faster the setting speed* and the *greater the temperature rise*. The temperature rises rapidly at the resin centre *after the resin has gelled*, which causes internal stress and leads to cracking on the outside. Once a crack starts on the outside it rapidly spreads inwards to relieve the build-up of stress within the material.

When making large castings, ways must be found for slowing down the speed of resin cure. This may be done by:

(1) reducing the volume of resin cast in one pour and increasing the number of layers cast;

(2) limiting the quantities of catalyst, so reducing the temerature range and extending the curing time;

(3) adding large quantities of fillers to reduce the resin content in the required volume.

On this page and the two following pages, steps are shown for the construction of a very large polyester sculpture. Clay is used to produced the basic figure sculpture (Figure 10.6A). Metal *shims* (Figure 10.6B) are then pressed into the clay around the split line. Pottery casting

plaster is mixed (see page 108) and applied as a thin coating over one half of the mould surface (Figure 10.6C). This area is then covered with jute *scrim* (Figure 10.6D), dipped in plaster, criss-crossed to form several interlocking layers, and left to dry. The other mould section is treated in the same way. When hard, the mould is carefully opened (Figure 10.6E), cleaned and left to dry, then sealed and waxed (Figure 10.6F). A wire former (armature) (Figure 10.6G) is prepared to fit inside the mould. This becomes the skeletal structure of the figure. While the mould dries a suitable filler is prepared — granite chippings, anthracite coal, or some similar cheap, inert mineral material (Figure 10.6H). A catalysed resin mix is prepared, and this filler or aggregate added (20% resin, 80% filler by weight), sufficient to fill the mould. The mould halves are packed with the mix around the wire armature (Figure 10.6I), joined and strapped firmly together (Figure 10.6J) and left to cure in a safe place (outside under cover). When cured, the mould is stripped off (Figure 10.6K), the sculpture checked and, if necessary, the split line filed and rubbed down with dampened wet-and-dry paper before being waxed with beeswax and burnished (Figure 10.6L–M). The structure will be strong, free from stress and the same colour as the filler powder or granite chippings.

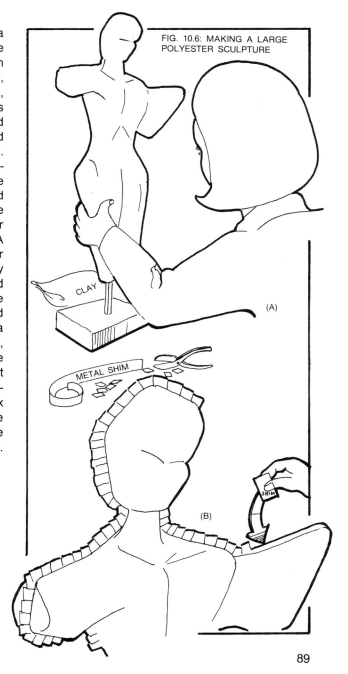

FIG. 10.6: MAKING A LARGE POLYESTER SCULPTURE

CLAY

(A)

METAL SHIM

SHIM

(B)

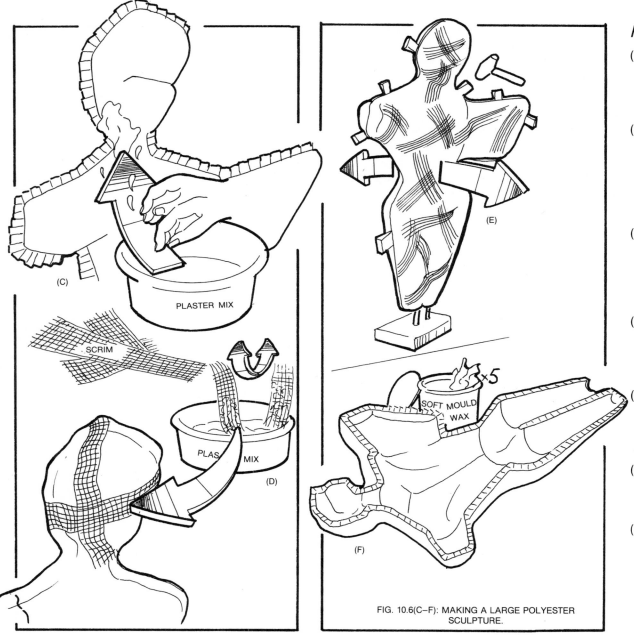

(C)

PLASTER MIX

SCRIM

PLAS. MIX

(D)

(E)

×5

SOFT MOULD WAX

(F)

FIG. 10.6(C–F): MAKING A LARGE POLYESTER
SCULPTURE.

PROJECTS AND ACTIVITIES

(1) Why is it necessary to limit the size of a resin casting or pour it in layers? What causes the resin to set and can you suggest why metal and mineral fillers make it possible to cast large mouldings in one go.

(2) Small castings often take a long time to cure fully. How can the rate of cure be speeded up? Devise an experiment to explore ways of making quantities of 3g weight cure rapidly. Then compare each specimen to see which is the most/least brittle. Make a chart showing your findings.

(3) An electronic timing device is needed for use on a motorcycle hill climb track. Could the electronic circuit be embedded in polyester resin? Would this be an advantage and what precautions would you need to take when making it?

(4) A friend has returned from holiday with some shells and dried seashore natural objects. Work out a casting plan to enable you to embed these items in resin to form a diorama.

(5) The biology department want some insect and plant specimens embedding so that they can be handled and displayed. In choosing the items, what potential problems must you avoid?

(6) Why must you pay particular attention to the problem of exotherm when preparing resin to make castings which will later be machine drilled?

(7) Using 5g amounts of polyester casting resin carry out some experiments with different percentages of different colour pigments to examine the effect of colour and quantity on setting time. Make a table of your results.

(8) Try experiment 7 but instead of using pigment make individual castings from each of the following: (a) plaster, (b) metal powder, (c) fine stone chippings, (d) wood flour and (e) coal dust. Make three castings with each material using 2, 3 and 4g of each. In making your observations examine each of the final castings to see that each has cured fully. Measure the mould and then measure each moulding and make up a chart to show the amount of shrinkage that has taken place in each sample.

(9) Can you find a commercially manufactured product made with polyester casting resin? Examine it closely and describe the type of mould that might have been used. If you cannot find an example of your own, describe how the moulds for the 'warrior' shaped chesspieces could be made.

(G)

(H)

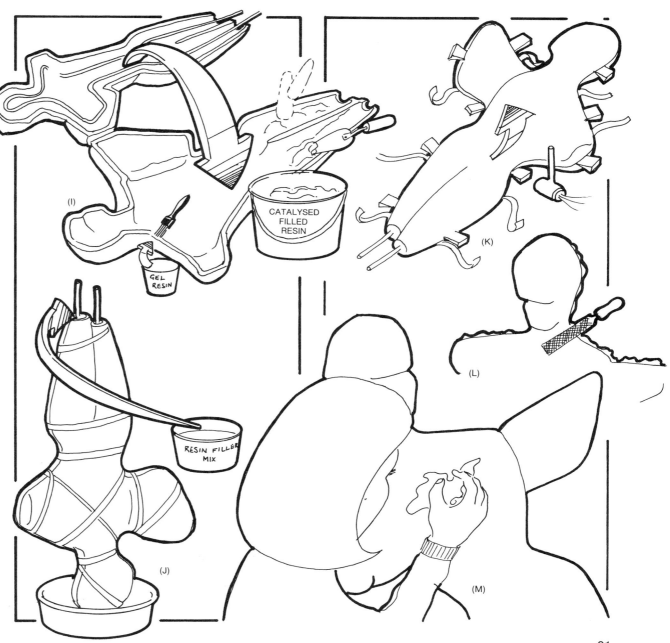

(I)

GEL RESIN

CATALYSED FILLED RESIN

(K)

(J)

RESIN FILLER MIX

(L)

(M)

Glass-reinforced polyester resin plastics are extremely strong and can be worked by hand into almost any shape. They have one smooth surface and one textured, and are resistant to water, weather and most chemicals. A variety of processes are used to make mouldings in both small and large quantities. Mouldings can be almost any size, ranging from boats down to products about the size of your hand.

THE MATERIALS

Glass fibre is a most important method of reinforcing many plastics materials, processed in a wide variety of ways. However, the term GRP has a very specific definition in schools and refers to products made by a variety of processes in glass reinforced polyester resins. The most common process is 'hand lay-up', which requires skilled people and is costly in time and wages. Other mechanized methods, such as spray lay-up, vacuum impregnation, and resin injection, are now widely used in industry with the same resins and glass fibre materials. Polyester resin mixed with glass fibre and slow curing agents are available for compression moulding as DMC (dough moulding compound) and SMC (sheet moulding compound), but are not within the scope of this book. Polyester resin by itself is rigid and brittle. Glass fibre on its own is flexible and has little structural strength. When these two are combined in a properly prepared moulding, an almost indestructible product results. Thin-skinned light-weight shell structures can be made that are tough, resilient, impact-resistant, and weather- and chemical-resistant. They can be translucent, or bright and colourful, suited to the most rigorous applications, and yet require only simple moulds. It is a unique and remarkable medium that does not need sophisticated machines. This makes it an ideal medium for prototyping designs. Where necessary, it can also be used to simulate plastics mouldings that would normally be commercially made in other materials by other processes, but which are not possible in schools. The size and complexity of a product is almost unlimited. The photographs here show some of the latest and most exciting applications where these unique materials have been used to their maximum advantage.

On the following pages, several methods are described for designing and making both products and moulds in GRP; all are within the scope of the average school pupil and workshop facilities. The resulting products will be highly professional, both technically and aesthetically, if the basic procedures are carefully followed.

Hand lay-up of GRP

Moulds, often called mould plugs, can be made from a wide variety of materials. Wood, medium-density fibre (MDF), hardboard and plaster are frequently used in schools. These materials are all porous and must be sealed with cellulose acetate sealer or shellac (shellac crystals dissolved in methylated spirits). Do not use household or car spray paints as they are often attacked chemically by the polyester resin.

Motor cyclist's crash helmet moulded from glass fibre reinforced polyester resin.

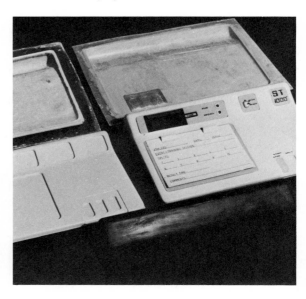

Electronic timer device moulded by hand lay-up in glass reinforced polyester resin.

Lotus car body moulded by injection of polyester resin into glass fibre. The main body is formed from two large mouldings.

Release agents are essential on all moulds except those of polyethylene, PVC, Vinamold, or wax. Resin manufacturers supply special non-silicone waxes or PVA (polyvinyl alcohol) and wax emulsion systems that are water-soluble. These should be thoroughly applied several times before moulding commences to ensure full separation and a top-quality surface to the finished product.

Polyester gelcoat

Once the mould has been prepared with release agent, a gelcoat can be applied. The gelcoat has several functions. It will form a tough hard-wearing surface, resistant to chemicals and the weather, as well as being able to withstand impact.

It is supplied as a thixotropic resin (a highly viscous fluid which thins when stirred rapidly) and requires pigment to colour it and a catalyst to make it cure. Fifty grams should be weighed out for every 1000 cm² of mould surface area. When mixed with the pigment and catalyst (Figure 11.1A) and brushed out over the mould, it will provide an even skin 1.5 mm thick (Figure 11.1B). It is usual to add up to 7% by weight of pigment. Use the catalyst according to the supplier's instructions — generally 1%, 2% or 4% depending on the system. The speed of cure of the gelcoat will depend on room temperature, pigment colour (yellow — fast; black — slow) and catalyst proportion. Thorough mixing is essential, followed by steady, even brushing-out over the entire mould surface. For mouldings of large surface area it is wise to catalyse small quantities of a previously prepared resin pigment mix. Catalysed bulk mixes of 1 kg or more cure too rapidly to be brushed out successfully. Once the gelcoat has been applied it must be left to cure for between 30 minutes and an hour. Brushes must be cleaned in solvent immediately. The cured resin will remain firm on the

mould if touched, but will feel sticky (this provides a chemical key for the next stage).

Warning: Use a barrier cream and wear gloves and goggles when measuring out and applying all resins. The MEKP (methyl ethyl ketone peroxide) catalyst must be kept in and dispensed from a labelled calibrated polyethylene measuring dispenser as this chemical is harmful to the skin and eyes. It should be used only under the direct supervision of your teacher.

Polyester laminating resin

Laminating or 'lay-up' resin is used to bond the glass fibre in layers onto the gelcoat. This resin is more 'watery' than gelcoat but requires the same MEKP catalyst to make it cure. The MEKP

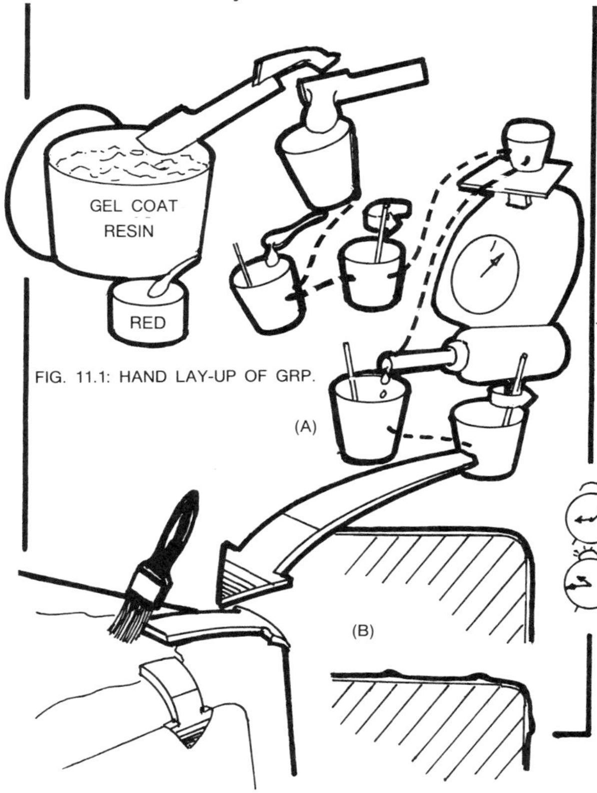

GEL COAT RESIN

RED

FIG. 11.1: HAND LAY-UP OF GRP.

(A)

(B)

FIG. 11.2A: PAINTING CURED GELCOAT WITH CATALYSED LAMINATING RESIN.

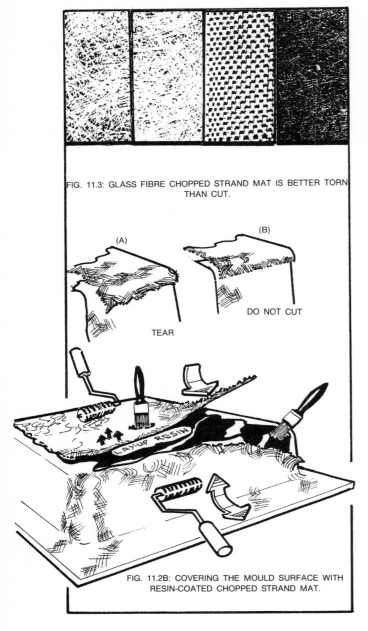

FIG. 11.3: GLASS FIBRE CHOPPED STRAND MAT IS BETTER TORN THAN CUT.

(A)

(B)

DO NOT CUT

TEAR

LAY-UP RESIN

FIG. 11.2B: COVERING THE MOULD SURFACE WITH RESIN-COATED CHOPPED STRAND MAT.

94

is used in the same proportions, but in this case cures over a longer time to enable the resin to be worked through several layers of glass mat. It is not essential to pigment this resin but it does give additional 'body' to the colour used in the gelcoat.

The procedure is simple. Resin is pigmented and then catalysed (Figure 11.2A). This resin mix is stippled into each layer of chopped strand mat (CSM) until each layer has been thoroughly impregnated. First, a coating of catalysed laminating resin is painted generously over the cured gelcoat. A piece of CSM is immediately placed over the wet resin and again stippled with short stabbing strokes of a brush. This action causes the resin to find its way up through the glass mat (Figure 11.2B). The entire mould surface is covered in this way, completing the first layer. While this is still wet, more resin is painted over the first layer, followed by another sheet of CSM, and then is stippled. Layer 2 is completed gradually, piece by piece. A roller is then worked *slowly* and methodically over the lay-up to bring all excess resin to the surface. Sufficient resin should be present to wet out a third layer of CSM. As one layer builds up on top of another, the pieces used should overlap at different points to achieve an even distribution of glass. A good lay-up has a pleasant texture but a fine surface tissue can be applied if desired. A formula for calculating resin/glass ratios is given on page 105.

Glass fibre

While the gelcoat is curing, the glass-fibre chopped strand mat can be tailored to suitable shapes, and in sufficient quantity to cover the mould surface in sections several times. It is best to tear the glass fibre to obtain a 'feathered' (Figure 11.3A) edge which is less noticeable than an edge cut with scissors (Figure 11.3B).

The number of layers will depend on the size of the moulding and its end usage. Glass-fibre CSM (chopped strand mat) comes in several weights (300 g/sq. m and 450 g/sq. m being the most suitable for school use). A moulding the size of a briefcase would require three layers of 450 g/m² CSM.

The glass from which the chopped strand mat is made is available in two main types, 'A' glass (Figure 11.4A) and 'E' (Figure 11.4B). The 'A' glass is an *alkali* glass, suitable for 'indoor' applications, while 'E' glass is a *borosilicate* or Pyrex-type glass, more suited to outdoor, marine, or electrical applications. Under an electron microscope a filament of 'A' glass looks like a solid clear rod, while a filament of 'E' glass appears hairy. In the latter sample these hairs combine chemically with the resin to make a better bond, making it more suitable for exterior use. There is no visible difference between either form of glass chopped strand mat.

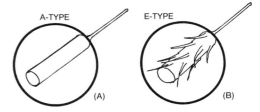

A-TYPE

E-TYPE

(A)

(B)

FIG. 11.4: A- AND E-TYPE FIBREGLASS, SEEN UNDER THE ELECTRON MICROSCOPE.

Each glass fibre is a bundle of filaments about 50 mm long. The mat is made up of a large number of straight fibres loosely held together by a starch bonding agent (silane). When making a moulding the silane dissolves in the resin, freeing the fibres and allowing them to curl and separate into their filaments. Catalysed laminating

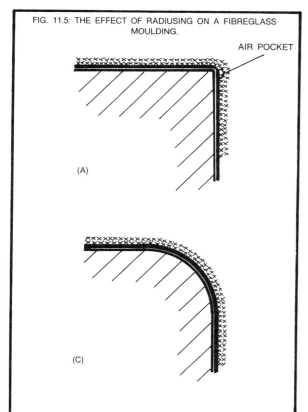

FIG. 11.5: THE EFFECT OF RADIUSING ON A FIBREGLASS MOULDING.

AIR POCKET

(A)

(C)

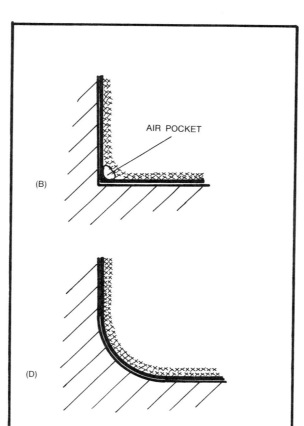

AIR POCKET

(B)

(D)

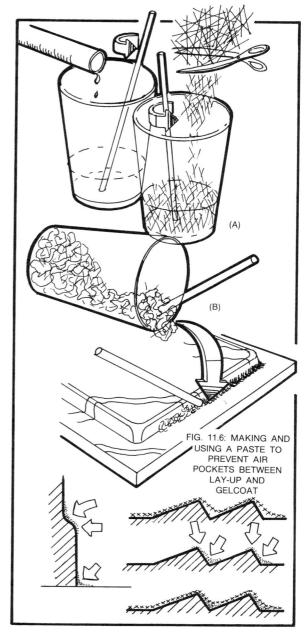

(A)

(B)

FIG. 11.6: MAKING AND USING A PASTE TO PREVENT AIR POCKETS BETWEEN LAY-UP AND GELCOAT

resin is worked up through the glass mat by a stippling action with a brush or gentle pressure from a roller. It is essential to achieve the right balance or ratio of resin to glass fibre. The fibres will break down into filaments and curl only when sufficient resin is present, and an excess of resin shows as a wet puddle hiding the fibres. When the glass fibres remain straight and have air spaces between them, the moulding is referred to as a 'dry lay-up'. A dry lay-up will be weak and absorb water.

Corners should be radiused on all fibreglass mouldings. The reason for this is that the glass fibre will not easily follow round sharp corners.

On positive corners the glass fibre bends outwards before following the side round, while in sharp negative corners the glass fibre jumps across. Figure 11.5A and B show these problems. A good radius (Figure 11.5C and D) makes corner moulding much easier. Often a gelcoat appears sound but has air pockets trapped behind it. When sharp corners are essential it is necessary to prepare a special paste from catalysed resin and finely cut-up glass mat (Figure 11.6A). This mix should then be applied with a stick — *not* a brush — into all difficult corners before continuing with the lay-up (Figure 11.6B).

95

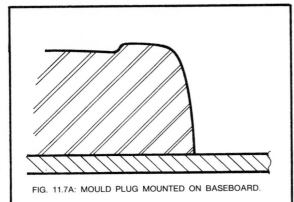

FIG. 11.7A: MOULD PLUG MOUNTED ON BASEBOARD.

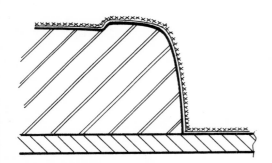

FIG. 11.7B: GLASS-FIBRE LAY-UP OVER PLUG AND BASEBOARD TO BE USED AS A NEGATIVE MOULD AT THE NEXT TWO STAGES (C AND E).

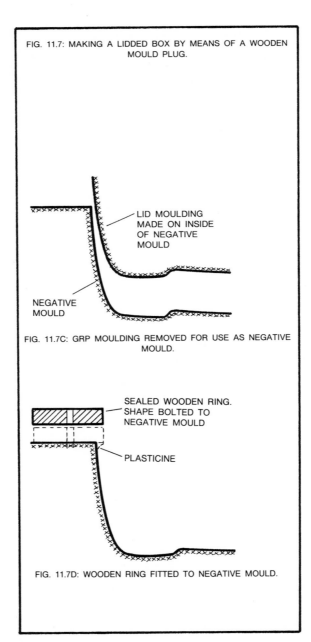

FIG. 11.7: MAKING A LIDDED BOX BY MEANS OF A WOODEN MOULD PLUG.

LID MOULDING MADE ON INSIDE OF NEGATIVE MOULD

NEGATIVE MOULD

FIG. 11.7C: GRP MOULDING REMOVED FOR USE AS NEGATIVE MOULD.

SEALED WOODEN RING. SHAPE BOLTED TO NEGATIVE MOULD

PLASTICINE

FIG. 11.7D: WOODEN RING FITTED TO NEGATIVE MOULD.

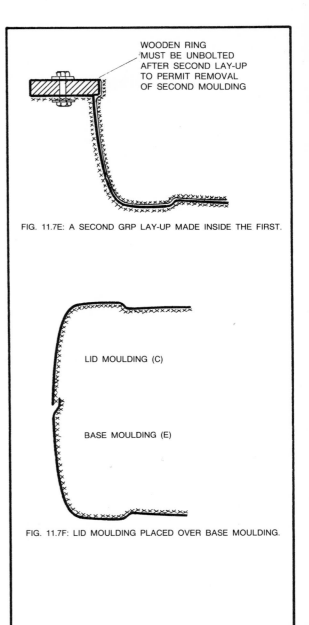

WOODEN RING MUST BE UNBOLTED AFTER SECOND LAY-UP TO PERMIT REMOVAL OF SECOND MOULDING

FIG. 11.7E: A SECOND GRP LAY-UP MADE INSIDE THE FIRST.

LID MOULDING (C)

BASE MOULDING (E)

FIG. 11.7F: LID MOULDING PLACED OVER BASE MOULDING.

LIDDED BOX

Figure 11.7 shows the essential steps for producing a box with a lid from a wooden mould plug.

A positive wooden mould plug is mounted on a baseboard already sealed with shellac and coated with wax release agent (Figure 11.7A). A glass fibre lay-up is prepared over the plug and over the baseboard. A gelcoat is first applied, allowed to cure, then followed by several layers of chopped strand mat and lay-up resin (Figure 11.7B).

The glass-fibre moulding is removed, waxed and polished and then used as a *negative* GRP mould (Figure 11.7C). *Inside* this mould another GRP moulding is produced, which will eventually become the box lid. A wooden ring is made to fit around the top of the negative GRP mould (Figure 11.7D). This is bolted onto the surround flange so that it projects 3 mm inside the mould. Plasticine is 'wiped' into the corner to form a radius between the inside edges of the mould and the wooden ring. A further GRP lay-up is made inside the mould so that it follows up the edges of the ring (Figure 11.7E). When this lay-up has cured, the ring is unbolted and the moulding removed. The lid moulding can now be fitted over the box base moulding (Figure 11.7F).

In this example the wooden ring creates an *undercut* or 're-entry' form, but because it can be removed easily it does not prevent the removal of the moulding. Careful mould design permits complex products to be made simply.

SPLIT MOULDS IN GRP

This yacht project shows how a split GRP mould is made. The yacht hull plug can be made from solid wood, solid plaster, or templates and stringers of wood. The plug must be dry and then sealed with shellac or cellulose acetate sealer, followed by several coats of wax release agent (Figure 11.8A)

The keel creates an undercut shape, making it necessary to form the mould in two parts which can be pulled off sideways (Figure 11.8B). To make sure that these two mould halves locate together accurately, it is necessary to put a false 'ribbon' flange made of hardboard around the plug. For exact location, buttons of Plasticine are moulded onto the good surface that has previously been sealed and waxed. Plasticine is also used to fill any minor gaps between this flange and the plug. The flange and half of the boat hull is then gelcoated and moulded in glass fibre and left to cure (figure 11.8C)

When fully cured this GRP moulding *is left in position on the plug* but the hardboard flange and all traces of Plasticine are removed (Figure 11.8D). The newly exposed flange and remaining half of the plug are waxed and laid up as before and left to cure (Figure 11.8E). Before separating the two mould halves, the flange is drilled with 6 mm diameter holes, every 60 mm, so that the two halves can be bolted together again (Figure 11.8F). The two negative mould halves now have to be laid up separately before being brought together (Figure 11.8G), because there is insufficient space to lay-up the keel area. To do this gelcoat is applied over each surface up to the point where the flange begins, and is then allowed to cure.

Glass fibre and laminating resin are applied over both moulds in the usual way, *except* for a 10 mm band just short of the flange. The gelcoat is left bare at this point.

When the resin has cured, the two halves of the mould are bolted together (Figure 11.8H). The area of gelcoat that is left exposed is then re-gelled and allowed to cure.

Finally, strips of chopped strand mat or woven tape are applied with laminating resin to cover both the gel-coated area and the surrounding glass fibre (Figure 11.8I). The two halves of the boat hull are now joined together as one moulding and left to cure.

The bolts from around the flange are removed (Figure 11.8J). Then chisel-ended strips of hardboard are used as wedges to separate the two flange faces and the mould halves from the moulding.

Finally the joint line is inspected, rubbed down with a fine wet-and-dry paper if necessary, and polished (Figure 11.8K). To fit a deck, small blocks of wood can be bonded with resin inside the top edge of the yacht hull moulding. Lead shot can be dropped into the keel space to make the yacht float to the right depth, but experiments under sail should be made before bonding the shot finally to the inside of the hull.

Split fibre glass mouldings for a garden stool showing the finished product.

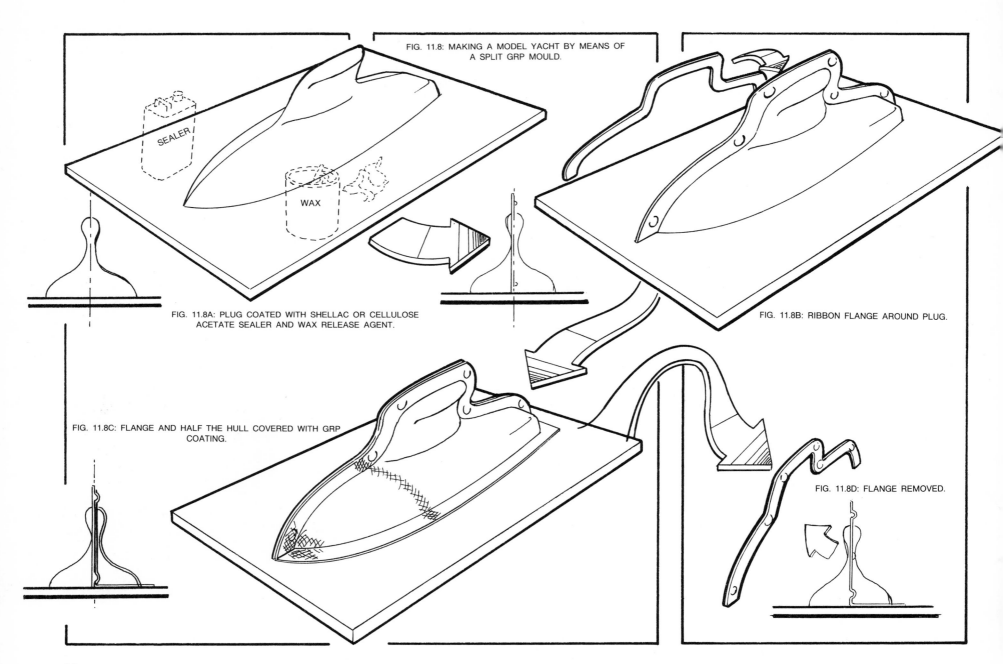

FIG. 11.8: MAKING A MODEL YACHT BY MEANS OF A SPLIT GRP MOULD.

SEALER

WAX

FIG. 11.8A: PLUG COATED WITH SHELLAC OR CELLULOSE ACETATE SEALER AND WAX RELEASE AGENT.

FIG. 11.8B: RIBBON FLANGE AROUND PLUG.

FIG. 11.8C: FLANGE AND HALF THE HULL COVERED WITH GRP COATING.

FIG. 11.8D: FLANGE REMOVED.

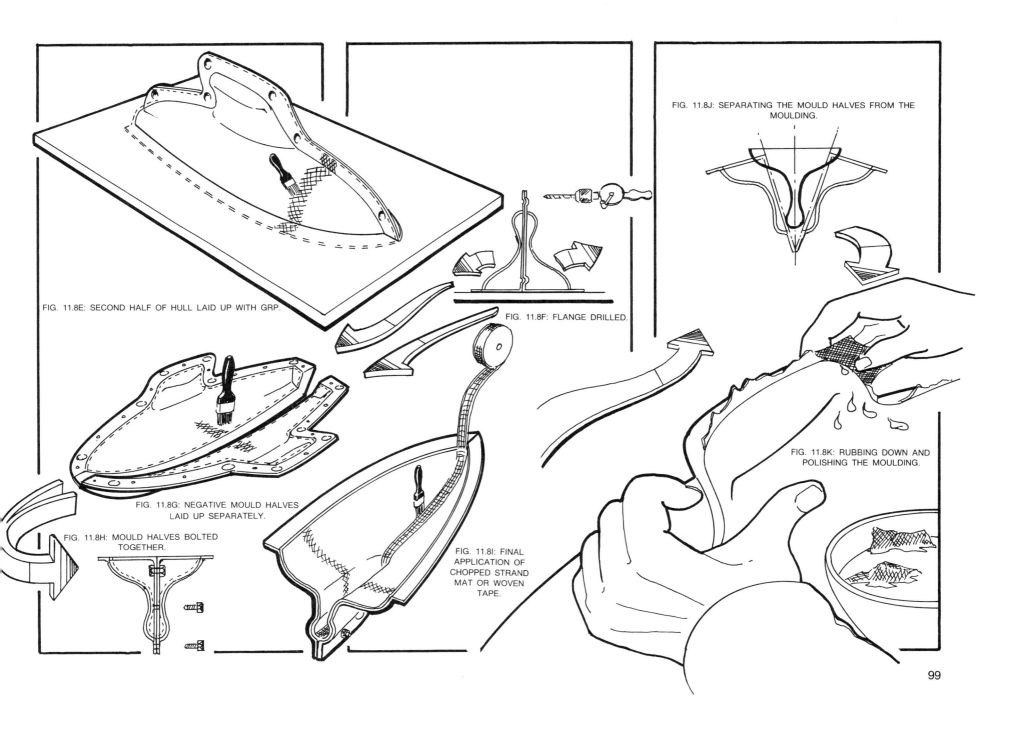

FIG. 11.8E: SECOND HALF OF HULL LAID UP WITH GRP.

FIG. 11.8F: FLANGE DRILLED.

FIG. 11.8G: NEGATIVE MOULD HALVES LAID UP SEPARATELY.

FIG. 11.8H: MOULD HALVES BOLTED TOGETHER.

FIG. 11.8I: FINAL APPLICATION OF CHOPPED STRAND MAT OR WOVEN TAPE.

FIG. 11.8J: SEPARATING THE MOULD HALVES FROM THE MOULDING.

FIG. 11.8K: RUBBING DOWN AND POLISHING THE MOULDING.

99

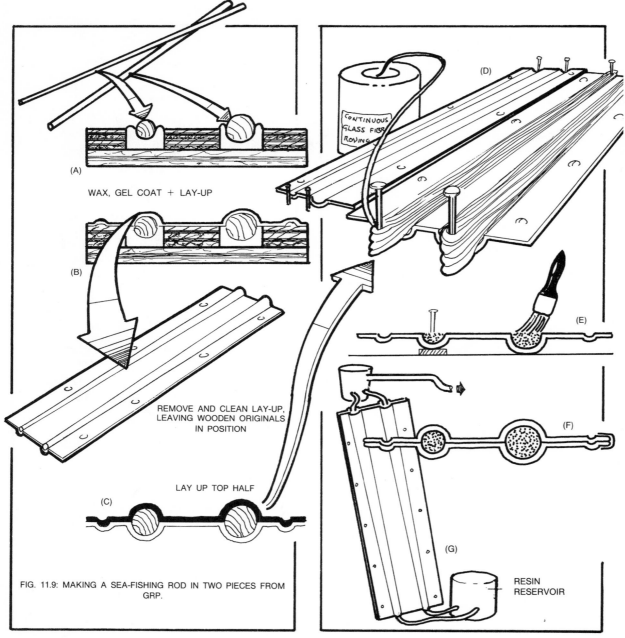

(A) WAX, GEL COAT + LAY-UP

(B)

(C) REMOVE AND CLEAN LAY-UP, LEAVING WOODEN ORIGINALS IN POSITION

LAY UP TOP HALF

(D) CONTINUOUS GLASS FIBRE ROVING

(E)

(F)

(G) RESIN RESERVOIR

FIG. 11.9: MAKING A SEA-FISHING ROD IN TWO PIECES FROM GRP.

FISHING ROD

Figure 11.9 shows how a two-piece solid glass fibre sea fishing rod can be made. Two 1m lengths of wooden dowel rod are tapered to match one another. These are then sealed, waxed and set in Plasticine in pre-cut grooves in a piece of plywood and mounted on a baseboard (Figure 11.9A). Care is necessary to ensure that the tapered dowels are set at exactly half their depth. Excess Plasticine is removed and made level with the ply surface. The ply is sealed and waxed. Small button shapes of Plasticine are spaced around the top outer surface of the ply to act as mould locators.

A GRP moulding is made in the usual way (Figure 11.9B). When cured it is removed from the surround but with the tapered dowels remaining in place. This moulding is cleaned down, all traces of Plasticine removed and waxed and a facing mould half laid-up in GRP as before (Figure 11.9C).

When cured, the split mould is opened, the dowels removed and everything re-waxed. The mould cavities of the two halves are gelcoated and left to cure.

A 'cheese' of continuous glass fibre roving (rather like coarse sisal string) is then run backwards and forwards along the length of each mould cavity (Figure 11.9D). A nail in the bench keeps the roving taut. With each pass, catalysed laminating resin is worked into the glass with a brush (Figure 11.9E). When all four cavities have been filled with glass and resin, the two mould halves are firmly bolted together and left to cure (Figure 11.9F).

It is possible to improve this process by adding a catalysed resin reservoir to feed one end of the mould and a resin trap and vacuum pump at the other (Figure 11.9G). This can be used to extract the air from the moulding. When the edges are firmly taped the system is closed.

Since air rises, the resin reservoir should be at the lowest point, the trap and vacuum pump at the highest level. Once the air has been evacuated the connecting tubes can be clamped and the resin left to cure.

WINDING

Short lengths of slightly tapered tube and cone forms can be made by this process (Figure 11.10).

A baseboard with two support stands is required, one to carry a roll of glass-fibre roving, the other to support the mould.

The cheese of glass-fibre roving is always unwound from the middle. Occasionally the coils unravel too quickly and become tangled. It is worth preparing a special roll of glass fibre sufficient to supply the moulding being made (Figure 11.10A). In this way the cheese of glass does not become contaminated with resin.

Between the two is the resin reservoir. A bar or roller is situated in the reservoir, which is topped up with a catalysed resin mix (Figure 11.10B). The glass-fibre roving passes into the resin trough under the roller and is wound directly onto the mould. As the mould revolves, the glass is drawn through the resin and guided up and down the length of the mould, each layer criss-crossing the previous one (Figure 11.10C). From time to time the resin excess needs to be brushed out over the surface.

For decorative objects such as the lampshade shown (Figure 11.10D), coloured tissue paper or printed paper patterns can be moulded-in between the layers of glass.

Provided you remember that the mould has to be removed from either one or both ends of the tube and must therefore be tapered, a wide variety of designs is possible.

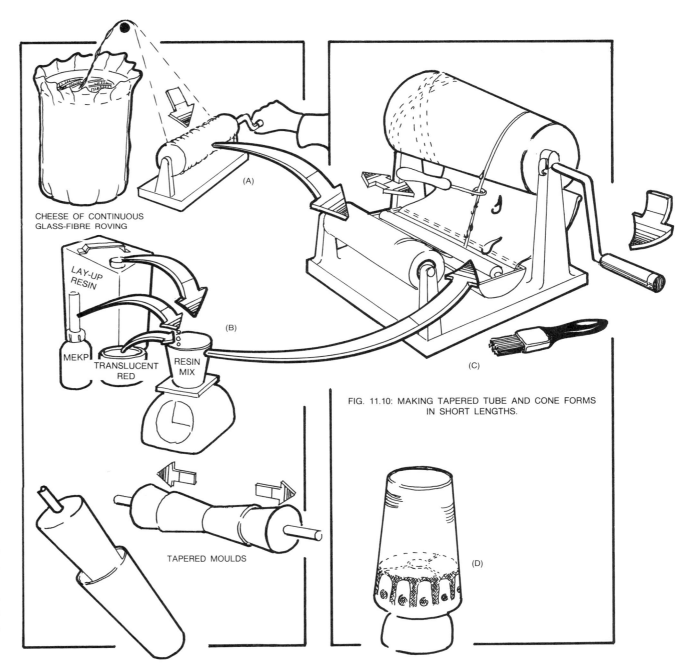

CHEESE OF CONTINUOUS GLASS-FIBRE ROVING

LAY-UP RESIN

MEKP

TRANSLUCENT RED

RESIN MIX

(A)

(B)

(C)

FIG. 11.10: MAKING TAPERED TUBE AND CONE FORMS IN SHORT LENGTHS.

TAPERED MOULDS

(D)

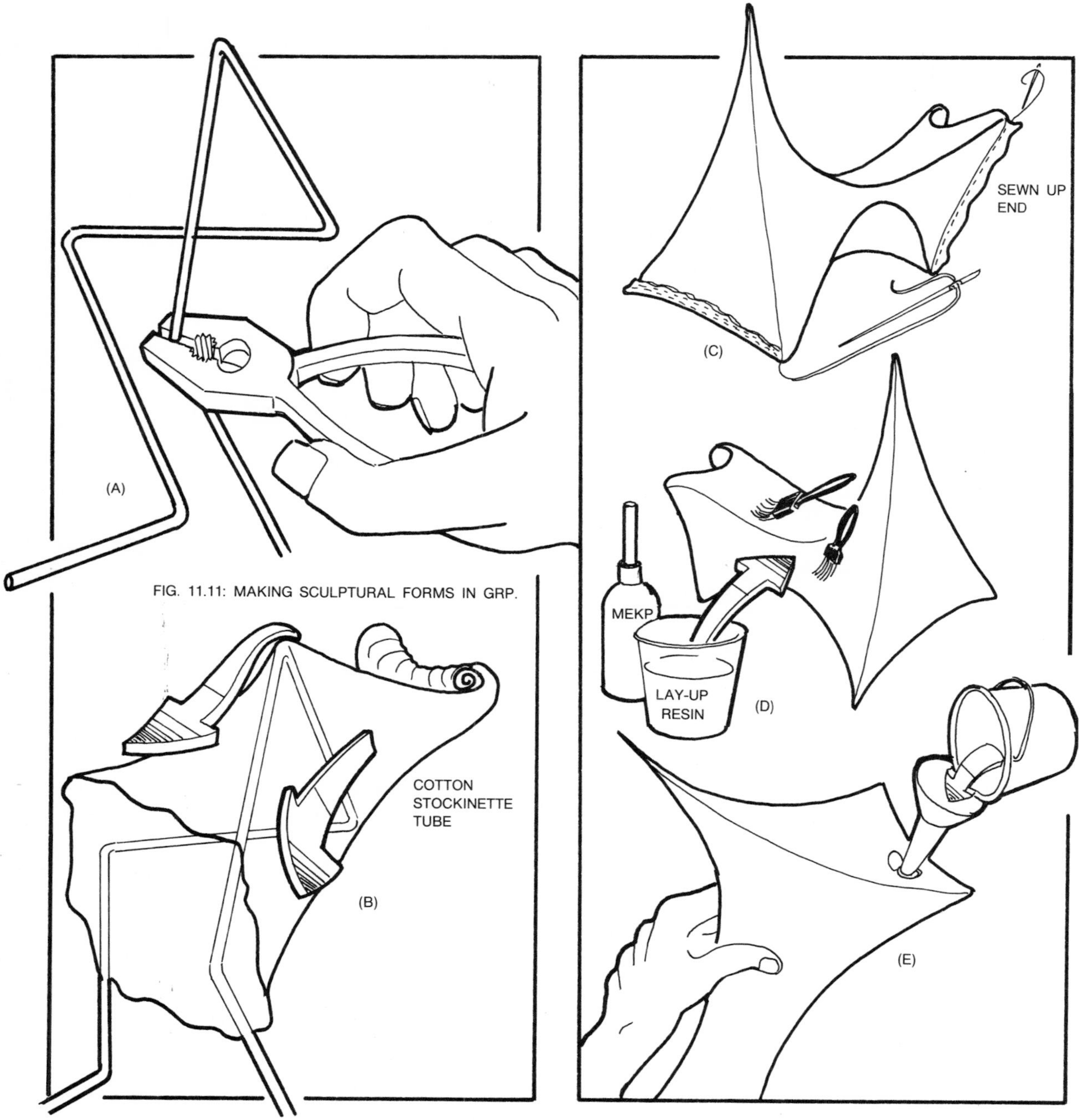

FIG. 11.11: MAKING SCULPTURAL FORMS IN GRP.

(A)

(B)
COTTON
STOCKINETTE
TUBE

(C)
SEWN UP
END

(D)
MEKP
LAY-UP
RESIN

(E)

SCULPTURE

Some interesting sculptural forms can be created in GRP without the use of a mould. It requires experimentation and planning to control the shapes and to get the best out of the technique. First a wire or steel rod armature is bent into an interesting 'space frame' form (Figure 11.11A). The ends can be wired, soldered or brazed together. When rigid, cotton stockinette is pulled over the shape (Figure 11.11B). This is a two-way stretch fabric and it stretches, curves and dips in the most pleasing of natural shapes. The tension can be adjusted in different areas and then the material can be lightly stitched to the frame making a complete enclosure (Figure 11.11C).

A mix of plain laminating resin and catalyst is lightly painted over the fabric, which moves and absorbs the resin (Figure 11.11D). At the end of the operation it may be necessary to adjust the tension slightly, especially on a large moulding. This should now cure fully for 24 hours.

The cured structure will be stiff but not as strong as a GRP moulding. The structure should be laid up with two or three layers of CSM, fully wetted out and then left to cure again. To make the sculpture solid, a hole is made in one side and the cured moulding filled with sand or plaster (Figure 11.11E). A glass-fibre patch is worked over the hole and feathered out and left to cure.

A mix of metal filler powder and catalysed resin is worked into a paste and trowelled over the surface. The 'trowelling action' will create a natural texture and should blend with the lines of movement in the form. When cured this surface can be burnished, 'patinated' and 'boot-blacked' to suggest weathered metal.

ADDITIONAL REINFORCEMENTS

Large GRP moulds and mouldings are often difficult to handle because of their flexibility, and sometimes it is necessary to make them more rigid without incurring the expense of several complete extra laminations.

Two well known methods of adding local ribbing are used. The first is carried out with paper rope (Figure 11.12), which is available in several diameters and comprises a soft iron wire core generously covered with a highly absorbent tissue paper, all bound together in a cotton matrix. The second method uses either rigid or flexible bars of polyurethane foam (Figure 11.13).

The procedure is to rub down the back of the moulding to be reinforced and apply a broad band of resin-impregnated glass-fibre CSM over the surface. While this is wet the paper rope or polyurethane foam is anchored temporarily into position. A second layer of resin and CSM is carefully worked over the rope or foam. The assembly is then left to cure.

The tubular rib now formed in the glass fibre adds great strength to the shape (Figure 11.13).

When large flat surfaces have to be achieved, areas of 'tempered' hardboard can be used in the same way. The hardboard has to be perforated with large holes to enable it to be bonded in at regular intervals (Figure 11.14).

It must be remembered that the resin and glass reinforcement will shrink as they cure. To avoid serious marks showing on the good gel-coat side of the moulding, these practices should not be carried out until the moulding is fully cured (after about one month).

When it is necessary to fix mouldings to other structures, 'big-head' bolts or similar fittings should be bonded into the moulding. These look like threaded pieces of perforated metal and are designed to spread the load over a large area (Figure 11.15).

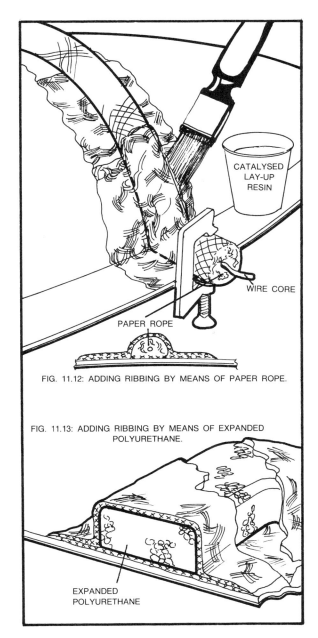

CATALYSED LAY-UP RESIN

WIRE CORE

PAPER ROPE

FIG. 11.12: ADDING RIBBING BY MEANS OF PAPER ROPE.

FIG. 11.13: ADDING RIBBING BY MEANS OF EXPANDED POLYURETHANE.

EXPANDED POLYURETHANE

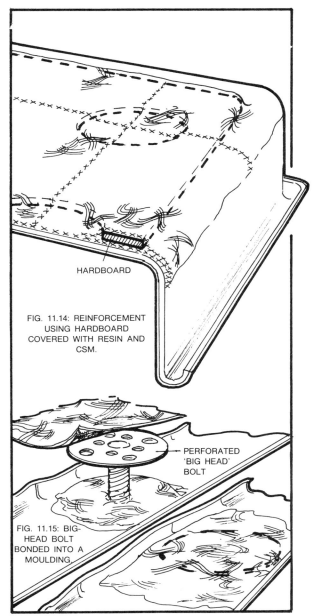

HARDBOARD

FIG. 11.14: REINFORCEMENT USING HARDBOARD COVERED WITH RESIN AND CSM.

PERFORATED 'BIG HEAD' BOLT

FIG. 11.15: BIG-HEAD BOLT BONDED INTO A MOULDING.

PROJECTS AND ACTIVITIES

(1) Can you find any examples of GRP products in your home? If so, describe them, paying particular attention to any important design features. If they have been in use for a long time describe any damage that is visible. Could the product be redesigned to avoid this sort of damage? Illustrate your answer with sketches.

(2) GRP is used to make many 'short-run' (limited production) products. Why is this?

(3) Collect illustrations that show GRP in use for transport applications. Why is it such a good material for this purpose? And, if you agree that it is good in this application, why are most cars made from steel?

(4) When designing products to be made from GRP what are the most important design features that have to be considered?

(5) *This experiment should be carried out in a fume cupboard.* Weigh out five samples, each of 50 g, of polyester resin into five paper cups. Add 1% catalyst to the first, 2% to the second, 5% to the third, 7% to the fourth and 10% to the fifth. Wrap a thin piece of cellophane around the bulb of each of five thermometers. Take time and temperature readings of each sample every few minutes. Note the time it takes for each sample to set (gel) and continue to measure the temperature after setting. Make a graph showing your results, marking in the gel point for each material.

(6) Repeat experiment 5 using *yellow pigment*, adding 7% pigment to each resin sample before the catalyst.

(7) Repeat experiment 6, with *black pigment*. Compare your results.

(8) Design a device that will allow a four-year-old child to carry things in the garden. It should provide a mechanical advantage that will permit it to carry a weight three times heavier than it would normally attempt.

(9) Is it possible to design a hydrofoil-type water ski that will enable a waterskier to stand 400 mm above the water surface? Would GRP be suitable for this? (This idea might provide a foundation for a new water sport.)

(10) Why are the following materials unsuitable for mould-making for GRP (1) Polystyrene (foam and sheet). (2) Chipboard. (3) Damp plaster. (4) Household oil-based paints.

RATIO OF MEKP (METHYL ETHYL KETONE PEROXIDE) CATALYST TO POLYESTER RESIN

Resin	1%	2%	4%
57 g	0.5 ml	1 ml	2 ml
114 g	1 ml	2 ml	4 ml
171 g	1.5 ml	3 ml	6 ml
227 g	2 ml	4 ml	8 ml
284 g	2.5 ml	5 ml	10 ml
341 g	3 ml	6 ml	12 ml
398 g	3.5 ml	7 ml	14 ml
455 g	4 ml	8 ml	16 ml

GELCOAT QUANTITIES

Gelcoat should be applied to the mould to give an even coating of 1 mm thickness. A useful approximation is:

45 g for every 300 mm²
or 455 g for every 1 m².

CSM (CHOPPED STRAND MAT) GLASS FIBRE

Glass fibre suitable for school use is available in two dry weights: 300 and 400 g/m². These may be used in any convenient combinations to make up the required amount of reinforcement. For example:

1 layer of 300 g/m²
can be added to: 2 layers of 450 g/m²
to give: 3 layers of 1200 g/m²

This combination would be suitable for a small chair, which would be strong but just able to flex when sat in. The choice is based on the function of the object and the rigidity and weight required. Strength is achieved by moulding shape, related to weight of glass fibre used. Mouldings of 0.1 m² surface area or less would require two layers of 300 g/m² glass mat: a large chair of 1 m² surface area would require three layers of 450 g/m² and one layer of 300 g/m² overall, with perhaps an additional layer of 450 g/m² where a central pedestal leg is mounted.

LAY-UP RESIN TO GLASS RATIO

Use the following formula to find out the total quantity of resin needed. This can be pigmented as a bulk quantity. Do not add catalyst to more than 450 g of coloured resin at a time. Larger quantities would set in the pot before they were used. Mix a little at a time, preferably 225 g, and work continuously, *following one mix with another* until the full lay-up is complete.

Formula: Weight of dry glass fibre × area × 2 (multiplication factor)
number of g/m² × number of m² × 2 = number of grams of resin

Thus, for a chair of 1 m² surface area:
3 layers × 450 g/m² = 1650 g/m² × 1 m² × 2 = 1650 × 2 = 3300 g resin

MOULD SEALER

(E) FROM THIS PLUG, A GRP NEGATIVE MOULD IS MADE. MOULDINGS ARE LAID UP INSIDE THIS. (SEE PAGE 96).

(A)

(B)

(C)

(D)

(F)

CAR BODY FILLER AND HARDENER

FIG. 12.1: STAGES IN MAKING A WOOD OR HARDBOARD MOULD FOR A SUITCASE.

WOOD, HARDBOARD CONSTRUCTION

The mould construction shown on this page is for a small suitcase. Figure 12.1A shows the basic framework of the case shape mounted on a baseboard. Joints are mitred where possible to hide the endgrain of the wooden members. The sides taper outwards and shaped stretcher rails provide support for the top hardboard panel. Fine battens of wood are mounted inside the frame and around the baseboard to support the countersunk hardboard top and side panels. This is important. Figures 12.1B–C show the difference between insetting the hardboard and mounting it on top of the frame. When inset the wood frame can be planed to make the corner radius without damaging the hardboard surface. When placed incorrectly on top of the frame a feather edge is created on the hardboard which is difficult to finish.

The frame must be a true rectangle as it sits on the baseboard; the diagonals between opposite corners must measure *exactly* the same distance. The slightest difference between these measurements will be exaggerated when the mouldings are eventually brought together face to face.

The hardboard panels are accurately cut to size and pinned into position. A 3 mm thick plywood surround is cut and fitted (Figure 12.1D). This plywood is partly sawn through several times to enable it to go round the corners. It can be shaped to provide support for hinges, locks and the handle, giving more strength to the edge.

The complete mould and baseboard should be sealed with cellulose acetate sealer or shellac (Figure 12.1E). Finally car-body filler should be used to fill pinholes, poor joints and sharp radii (Figure 12.1F). It should then be resealed, rubbed down, waxed and polished.

OTHER MOULD-MAKING MATERIALS FOR GRP PRODUCTS

A wide variety of materials is suitable for making moulds. One must be selected to provide the best method of construction and to achieve form, strength, accuracy and surface finish quickly.

Small mould plugs can be made in fine-grain solid wood, clay, plaster, wax or Vinamold.

Large moulds can be built on wooden frames or steel and dexion structures, and then clad in plywood, wood veneers, melamine-faced board, hardboard, sheet tinplate or aluminium, acrylic sheet or even GRP. Large multicurved shapes can be built up in chicken wire, paper and pottery casting plaster over wood frames, as shown on the following pages.

Wood

Ideal for many mould-making applications, provided it is dry and free from shakes (splits), and has fine grain. It can be used both for the structure and (in plywood form) for surfaces. All grain pattern and exposed joints must be filled and sealed, otherwise they will show on the mouldings. Medium density fibre (MDF) is an ideal mould-making material.

Hardboard

An excellent material, capable of shallow three-dimensional curvatures, generally used for surfacing over a wood frame structure. Tempered (waterproof-grade) hardboard is better than standard-grade. When using hardboard, avoid damaging the smooth surface.

Pottery casting plaster

An excellent medium, which can be cast or shaped by several methods (see following pages). Remember that any plaster construction must be thoroughly dried out before it is sealed and prepared for GRP work.

Thermoplastic sheet materials

Acrylic (PMMA), CAB, polyethylene and PVC sheet can be used flat or as moulded shapes. Polystyrene sheet and ABS sheet are both affected by polyester resins and should *not* be used for this purpose.

Thermosetting plastics sheet

The various brands of melamine-faced and polyester-faced boards are all suitable.

Foams

Rigid polyurethane foam is unaffected by polyester resin and can be surfaced with laminating resin and tissue, but requires considerable effort and time to achieve a good finish if 'painted' with polyester resin. Polystyrene will melt.

Glass

Sheet glass can be used if essential. Most of the textured grades are available in Perspex, which is less dangerous.

Sheet metal

The most common sheet metals used in mould-making for GRP work are tinplate and aluminium. As flexible media they are often used to obtain two-dimensional curvatures.

Polyester resins and car body fillers

These are used to surface-coat and fill difficult parts on a plug — radii, grain and imperfect joints. Mix and use a little at a time. When rubbing down to obtain a good surface finish use dampened wet-and-dry abrasive papers.

Vinamold

This is a flexible PVC material that is melted and poured over small complex shapes to make accurate moulds. When cold the material is 'peeled off' the original model. This material can be used to reproduce many small polyester resin castings and glass-reinforced mouldings, after which it may be melted down and recast into another mould form. The material does not need sealing or coating with release agents. See also pages 80 and 81.

Textures

Textured patterns are sometimes desirable to add interest and relieve large plain areas. Local areas of textured wallpaper, glasspaper, fabrics and rubber mouldings can be let into plugs. Even thin tape and transfer letters 'print' and eventually come out 'in relief' on the finished product. Beware, they can come out in reverse!

Silicone rubbers and flexible polyurethane compounds

These are thermosetting compounds that, once mixed, cast and cured, can only be used for making one mould. They reproduce very fine detail and are very tough and durable, but are generally too expensive for school use. There are health hazards attached to the handling of these chemicals, and manufacturer's literature must be studied and followed closely. The final cured material will withstand much higher temperatures than Vinamold.

Natural rubber: Latex

An excellent flexible mould-making material for use with polyester casting resin. It is applied to the original positive former by dip-coating, or brushing on in layers. Highly elastic; strips off complicated moulds easily. It can be made more durable by incorporating a butter muslin or cotton scrim within the brushed-on surface.

Wax

Solid blocks of candle wax can be cast and carved easily. A candle flame may be used to flame-polish carved or scratched surfaces. Usually used for making lost-wax jewellery, which, instead of being cast in metal, may be cast in polyester resin. Delicate wax shapes can safely be stored in a water bath without fear of damage (see page 86).

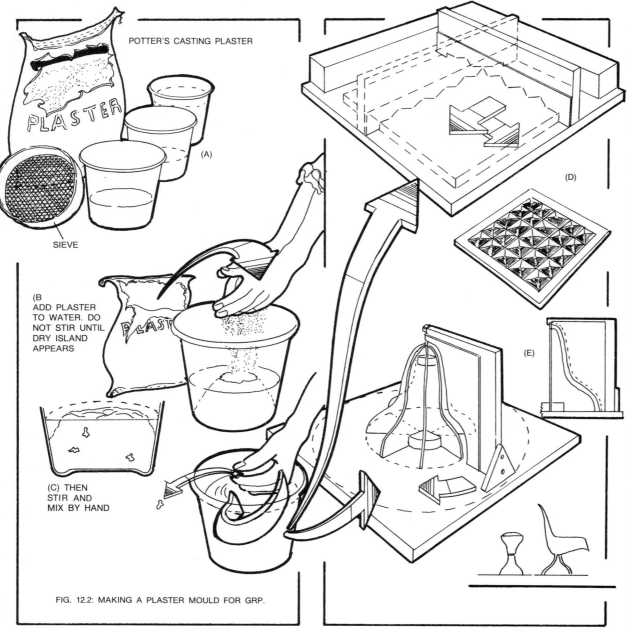

POTTER'S CASTING PLASTER

SIEVE

(A)

(B ADD PLASTER TO WATER. DO NOT STIR UNTIL DRY ISLAND APPEARS

(C) THEN STIR AND MIX BY HAND

(D)

(E)

FIG. 12.2: MAKING A PLASTER MOULD FOR GRP.

PLASTER MOULD-MAKING FOR GRP

The methods shown on these pages convey the principal techniques for shaping wet plaster using templates. The process is called *strickling* and is suitable for 2D and 3D shapes, and for shaping large structures.

Mixing

Use pottery casting plaster and add plaster to water by hand (Figure 12.2B). Do not stir the mixture at this stage or get your hands wet. Remove any lumps from the dry powder. The plaster will slowly build up in the bucket to create an island (Figure 12.2C). Keep adding plaster until the water is completely saturated with plaster and the surface is a low dry island. Allow to stand for two minutes. Then stir with your hand. Once stirred, the plaster will start its chemical reaction with the water and will set hard in about 15 minutes. So use the plaster as soon as possible after it has been stirred.

For thin mixes to top-coat and surface-finish, add plaster to water until an island is formed that is half the area of the water surface.

A similar watery mix can be used for dip-coating newspaper strips (100 mm wide by 400 mm long), to cover chicken wire or be used between templates. Before adding a fresh plaster mix to a dry plaster or wood surface, always dampen the area first.

Always stir the mix with your hand and feel for lumps, which must be removed. *Keep a second bucket of water nearby* to wash your hands and templates. *Do not put any plaster water down the sink. Do not scrape dry plaster out of new buckets — wait until hard and then flex the bucket, when everything will come away cleanly.* For big moulds a system of three buckets is sensible (Figure 12.2A): fresh mix; remains of previous mix, hardening off; washing water.

108

Strickling (two-dimensional)

Textured tile patterns can be achieved by using a template made of galvanized steel sheet pulled over a wet plaster mix in two directions, first one way then the other (Figure 12.2D).

A plywood baseboard, well greased, fitted with two straight-edged battens at 90° to one another, make the working platform. The shaped edge of the template is sharpened and the template fixed to a piece of plywood for rigidity. The template must be held vertically and must always travel across the plaster in the same direction, so that excess material collects only on one side. Several plaster mixes will be needed, the last being more watery. A wide variety of interesting patterns and tile shapes can be made in this way.

Strickling (three-dimensional)

The same principles apply, except that the template must revolve around a central point. The inner shape can be outlined with hardboard templates and any large spaces filled with crumpled newspaper (Figure 12.2E).

MAKING A LARGE CHAIR FORM

A structure is made of hardboard, chipboard or plywood templates, mounted on a strong baseboard (Figure 12.3A). Spaces are packed with newspaper and covered with chicken wire or builders' jute scrim. The plaster is applied and worked with straight-sided 2mm-thick strips of PVC sheet (Figure 12.3B). Curvatures are achieved by bending the sheet. Sharp curves need specially made templates (Figure 12.3C). It is often worthwhile to set pins into the template edges to control the build-up of plaster thickness. Plaster can also be cast or applied as a thin skin over clay, as shown on page 90. Plaster must be thoroughly dried before sealing, waxing and moulding in GRP.

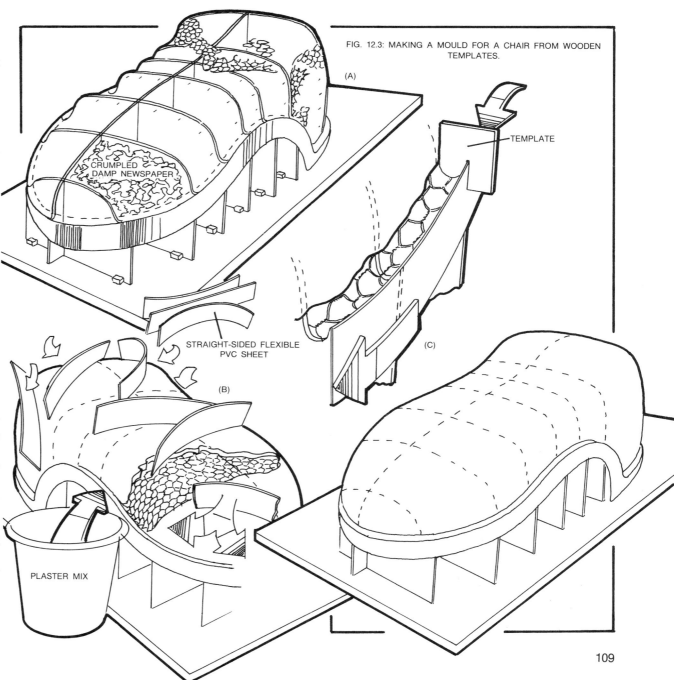

FIG. 12.3: MAKING A MOULD FOR A CHAIR FROM WOODEN TEMPLATES.

(A)

TEMPLATE

STRAIGHT-SIDED FLEXIBLE PVC SHEET

(B)

(C)

CRUMPLED DAMP NEWSPAPER

PLASTER MIX

PROJECTS AND ACTIVITIES

(1) Write a short description that describes the difference between positive and negative moulds. Do *not* use a diagram.

(2) When making glass fibre reinforced polyester mouldings you are advised to avoid sharp corners on the mould. Why is this? Draw a diagram to show what happens to the materials on sharp corners.

(3) What is the main function of a gel coat and what precautions should be taken when applying it? Gel coat is said to be **thixotropic**. Find out what this term means.

(4) When a mould has been made in wood or plaster, it has to be sealed and treated with a release agent. Why is this and what materials should be used for moulds that will later be used for glass fibre reinforced polyester work?

(5) Plain highly-polished glass-reinforced polyester mouldings often appear visually boring or poorly proportioned. Carry out some experiments using mould materials of different textures (for example wallpaper, transfer letters and strips of sellotape) to see if there are any opportunities to make 'glass fibre' mouldings more interesting. If you have time, experiment with masking techniques and several different colour gel coat applications. Make a wall panel that shows both the original 'mould materials' and the finished moulding. Take care to seal and wax textured materials very thoroughly before making mouldings.

(6) What is a split mould? Invent a situation when you would choose to make a split mould and describe why you feel it necessary. What function do location bosses provide on the split-line flange?

(7) Vinamold is the trade name for a flexible moulding compound. When and how should it be used? Can it be re-used? How many grades are there, how do they differ and do they need release agents at any time?

(8) You have just vacuum formed two tray mouldings, one in polystyrene and one in CAB sheet (cellulose acetate butyrate). Neither is strong enough for your purposes and so you decide to make a glass reinforced plastics version. One of the two mouldings could be used as a mould, the other would dissolve into a sticky mess. Which of the two vacuum formings would you use?

13 MOULDING THERMOSETTING PLASTICS

COMPRESSION MOULDING

This is the oldest of the commercial plastics moulding processes, and is used to make products and components from thermosetting materials.

In all these cases heat and considerable pressure are used to change the material's form and chemical structure. Unlike thermoplastics, these materials cannot be reheated and reshaped, because a chemical change has taken place.

A weighed amount of thermosetting material in either powder or granule form is placed in a mould cavity. Sometimes lightly compressed powder in the form of pellets of measured quantity are used, and when this happens each pellet is preheated before moulding (Figure 13.1A).

The two main mould halves are fixed to heater plates and mounted on plattens. One platten is fixed, while the other is free to move, driven by a hydraulic ram. Machine designs vary: some have a moving top platten, while in others it is the bottom one that moves. In (Figure 13.1B) the bottom platten is the moving one. On large commercial machines, several hundred tonnes pressure can be applied.

The moulds are made of steel, the main cavity almost always being in the bottom half to contain the material charge. The positive part of the top half is smaller than the negative cavity by an amount equivalent to the wall thickness of the product. The two mould halves touch only at the *pinch-off* point (Figure 13.1C), a ridge running round the cavity. When the mould halves close together firmly, the thermosetting material flows rapidly round the cavity (Figure 13.1D), any excess escaping at this point. The excess can then lie in the narrow space between the outer faces of the mould halves. After two or three minutes the mould is opened and the moulding is pushed out by ejector pins (not shown). The

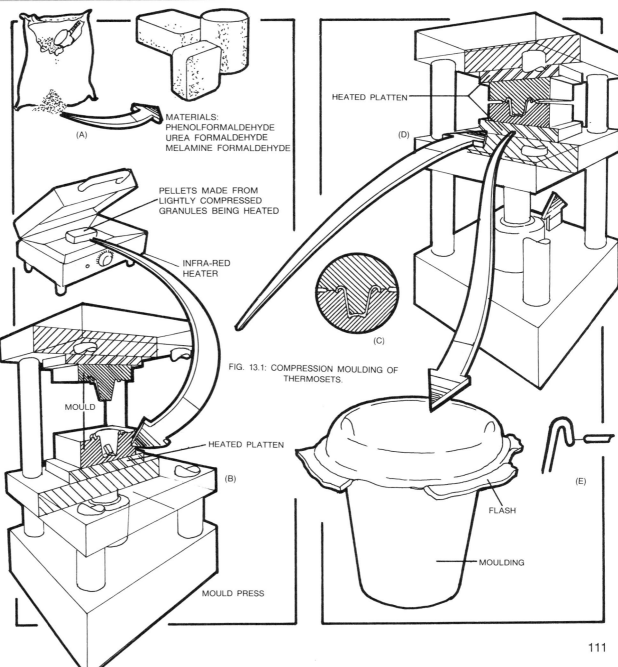

(A)

MATERIALS:
PHENOLFORMALDEHYDE
UREA FORMALDEHYDE
MELAMINE FORMALDEHYDE

PELLETS MADE FROM LIGHTLY COMPRESSED GRANULES BEING HEATED

INFRA-RED HEATER

MOULD

HEATED PLATTEN

(B)

MOULD PRESS

(C)

FIG. 13.1: COMPRESSION MOULDING OF THERMOSETS.

HEATED PLATTEN

(D)

FLASH

MOULDING

(E)

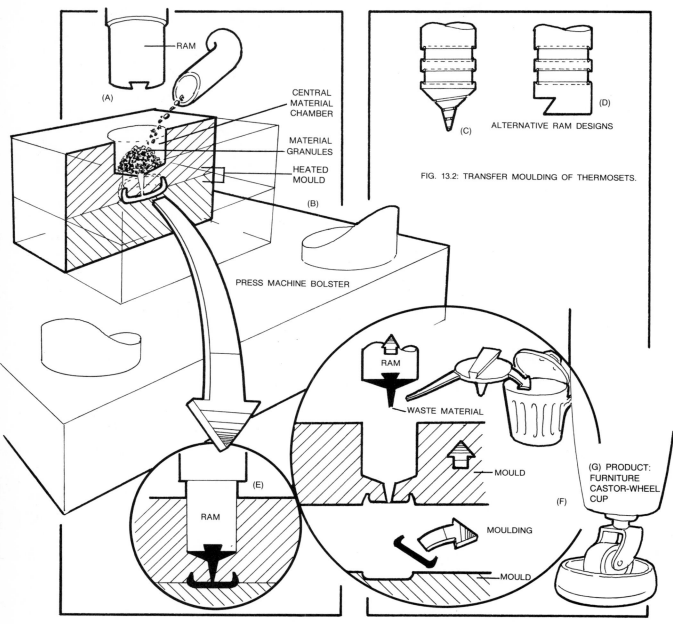

FIG. 13.2: TRANSFER MOULDING OF THERMOSETS.

flash (excess material) is broken off and discarded as waste, the flashline is cleaned, and the product is inspected. (Figure 13.1E). The mould is recharged for another cycle. The process is slow compared to injection moulding.

Transfer moulding

Transfer moulding is a form of compression moulding in which a charge of thermosetting material can be 'transferred' from one central chamber along runners to individual cavities. This system gives better mixing and curing characteristics.

The figures on this page show the central material chamber (Figure 13.2A), a shaped piston ram used to compress the material (Figure 13.2B–D), a sprue or runner, and the mould cavity (Figure 13.2E). The ram is designed to fit the central material chamber loosely (this is because in production the mould is heated and the ram is not; when not in production, both are cold and the ram will not fit). A very loose fit would cause loss of pressure during compression, so the ram collects some moulded material around itself, which acts like car piston rings.

The bottom of the ram is shaped so that material left in the central chamber in a cured state is lifted out automatically after each operation (Figure 13.2F). This is waste and is thrown away. Several end designs are shown (screw, wedge and tapered keyway). When the pressure is off, the ram retracts, the mould opens and the moulding is ejected from the cavity. The benefits of the process lie in the use of multiple mould cavities, all fed from a central material charge.

Some thermosetting plastic compounds can be injection-moulded on special machines but these are beyond the scope of this book.

DECORATING PLASTICS

The surfaces of most plastics articles are plain when they come from the moulding machine. They can be decorated in a variety of ways. Screen printing is used for decorating polythene bags and similar film materials, hot foil blocking is used to decorate small areas of thin PVC film and rigid thermoplastics mouldings, paint-spraying is used to decorate small areas of dolls' faces, and some large products are completely sprayed. Metal coatings can be applied to some plastics by electroplating or vacuum metallizing.

Hot blocking

The process of *hot blocking* is used to print small areas on thermoplastics mouldings, leather, paper, card and some painted surfaces. For thermoplastics products it is an ideal method for decorating small areas with type symbols and product names.

A hot die (Figure 13.3A) is used to press a metallic coated or plastic coloured foil skin onto the moulding (Figure 13.3B). The heat from the die melts the thermoplastics moulding where the letters or symbols press onto the surface. The molten plastic causes the coloured or metallic foil skin to stick where it touches, leaving the lettering or symbol coated and slightly impressed into the suface (Figure 13.2C). The most common examples of this can be found on pens and pencils, where the lettering is a different colour from the main body moulding. The technique greatly enhances the visual appearance of many of our everyday products, besides providing important identification of size, make and other data.

A small machine for use in school, designed to work on standard rolls of foil of common thermoplastics would make an excellent project.

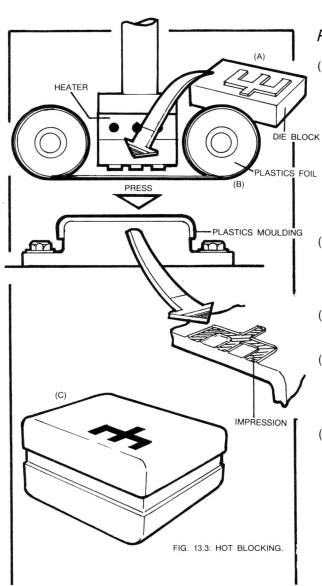

FIG. 13.3: HOT BLOCKING.

This school-made machine would not be suitable for use on card and other non-thermoplastics as high pressures are required.

PROJECTS AND ACTIVITIES

(1) Compression moulding is the oldest mass-production plastics process. A wide variety of early plastics products can be found in junk shops and at jumble sales. Start your own collection; it could be very valuable in a few years' time. Old radios and a wide range of domestic products from the turn of the century are still around and are often unwanted. (a) Trace their history, (b) take the trouble to find out when they were made and (c) follow their design changes up to the present time.

(2) Irons and kettles from the 1950s to 1970s had bakelite heat-resistant handles. Find some examples and draw them. What are today's products of this type made from? Can you give any reason for the change?

(3) Find examples of urea formaldehyde or melamine formaldehyde products in your home. List and describe them in detail.

(4) Find any examples of thermosetting compounds used in a car engine (pre-1980). What do you conclude from this? What changes have taken place since 1980 that have changed this?

(5) *Wear gloves and goggles for this experiment.* Obtain some black bottle tops. It is most probable that these are made from phenol formaldehyde. Take some lemon juice, vinegar, water, milk, car oil, nail-varnish remover and paint stripper. Immerse each top in one of these fluids for 24 hours, having weighed and recorded it first. Weigh again at the end of 24 hours and then re-immerse it for a week. Weigh each for a third and final time and compare the pattern of records throughout the experiment. Is phenol formaldehyde a good material for bottle tops?

113

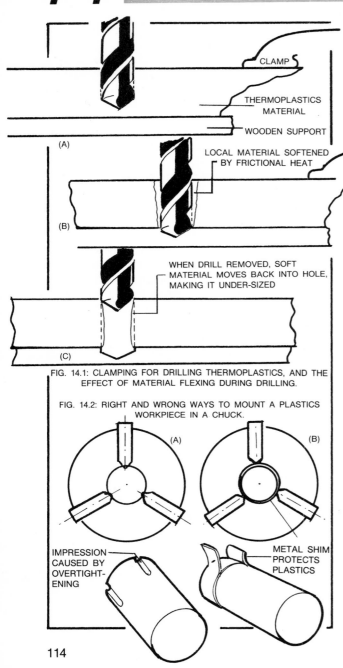

FIG. 14.1: CLAMPING FOR DRILLING THERMOPLASTICS, AND THE EFFECT OF MATERIAL FLEXING DURING DRILLING.

FIG. 14.2: RIGHT AND WRONG WAYS TO MOUNT A PLASTICS WORKPIECE IN A CHUCK.

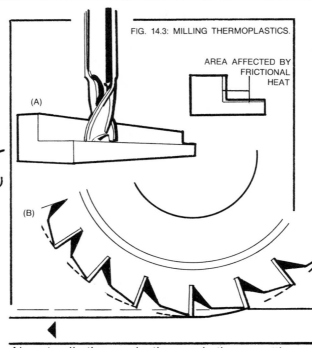

FIG. 14.3: MILLING THERMOPLASTICS.

Almost all thermoplastics and thermosetting plastics can be cut, drilled and machined by all the normal engineering machine processes. It is often necessary to alter the drill or cutter shape to suit the plastics material, but the tools must always be kept sharp.

Thermoplastics are poor conductors of heat, a fact that affects many machining operations. The temperature of the cutting tool, the material and the swarf must be kept below the melting point of the plastic. Softened plastics will blunt any cutting edge and increase frictional heat. It is usual to use a lubricant (thick soap solution, paraffin, soluble oil) to protect the plastics workpiece. Tools may be treated with anti-static agents to reject swarf. Softer plastics materials, such as polyethylene, will often be moved back by the pressure from a drill or milling cutter, especially if they become warm (Figure 14.1).

When the tool is removed, the hole or slot may be undersized as a result. To avoid this, the material can be chilled in a fridge before machining. All tools must have a good clearance angle so that swarf clears quickly.

CLAMPING

All materials must be held firmly throughout the machining operation. Sheet materials need to be fully supported, especially during drilling and sawing operations. Many plastics are said to be 'notch-prone' — that is, if a small crack starts it will 'run' when under pressure. Double-sided adhesive tape can be used to stick the material onto a wooden support block during drilling.

When holding soft or waxy material in a lathe chuck, it can distort or slip under jaw pressure. Soft metal strips called shims can both protect the material and distribute the pressure on it, and give added grip (Figure 14.2). Check from time to time that the material has not eased forward in the chuck.

SAWING

Very thin material can often be scored and then broken along the score line. Hand-sawing is suitable for most plastics materials, and ordinary woodworking tenon saws or coarse-toothed hacksaws work well. A coping saw or fretsaw with a coarse blade can be used to cut small intricate shapes; however, as the blades are short they often overheat, causing material to melt in the cut and weld together again after the blade has passed. Start by putting liquid soap along the cutting line so that the blade is constantly lubricated.

Small bandsaws are the best method for cutting complicated shapes, because the blade is long and keeps cool. The blades should have a wide setting and five or six teeth for every

10 mm (15 TPI). Senior pupils should use a band-saw only with the permission and under the direct supervision of the teacher.

Warning: All circular sawing must be carried out only by your teacher, using a special hollow-ground tipped blade. Advice on suitable saw blades and procedures is always available from materials suppliers.

When a material has been sawn it is often left with saw-tooth marks along the cut edge. To remove these marks use a file, scraper and fine abrasive papers dampened with water. The edge can then be polished using fine cutting pastes, or metal polish applied with a soft cloth. Edge polishing can also be carried out on a buffing machine using fine calico mops, but thermoplastics materials will often melt due to the frictional heat developed. The material must be fully supported and moved around during this operation so that localized areas do not become overheated. Hold the work firmly and apply only light pressure.

DRILLING

Standard high-speed steel drills can be used for holes up to 25 mm in diameter, but the chisel edge should be slightly off-centre to prevent undersized holes from being produced. Slow spiral drills give a better-quality finish than standard drills. Always use a coolant and constantly remove the drill. *Clear swarf from the drill and hole only when the machine is stopped.* A high speed for small drills and a slow, gentle feed is essential to make a smooth, clean hole. Slower speeds must be used for holes larger than 12 mm in diameter.

For polystyrene, acrylics, CAB, PTFE and PVC the drill tip should have a point angle of 118–130°. Nylons require a sharper point of 90°.

The cutting edges should have zero rake. The outside edges of the drill should penetrate the top surface of sheet material before the chisel edge has penetrated the underside. When drilling acrylic or brittle materials it is a sensible precaution to cover the surface with clear adhesive tape. This increases the surface strength and prevents cracks from developing. Always fully support the plastics material with a wood backing and clamp the material firmly.

When 25 mm or larger-diameter holes have to be made in sheet plastics, use a *hole-cutter* (Figure 14.4B), which is an adjustable blade mounted in a special holder, or a hole saw (Figure 14.4C). A hole saw is a special curved hacksaw-type blade mounted in a circular holder. Both of these drilling devices require a smaller pilot hole before they can be used. A slow speed and a slow feed are important.

Drilling generates more heat than any other machining process, making it essential to keep the work cool, especially on thermoplastics materials, which tend to soften and be pushed back by the drill, and then spring back when the drill is removed, making an undersized hole.

Warning: Always clamp sheet materials down for drilling. never hold them in your hand.

It is sometimes necessary to cast a block of polyester casting resin and then drill or machine it later. In this situation it is essential that the casting cures very, very slowly. The reason is that during cure the temperature rises, owing to the chemical reaction. The resin always sets before the temperature rises, and the chemical reaction starts in the centre of the block. The material tries to expand with the increase in temperature, causing considerable stress in some castings. When drilled or machined, the casting can shatter as the tool enters the

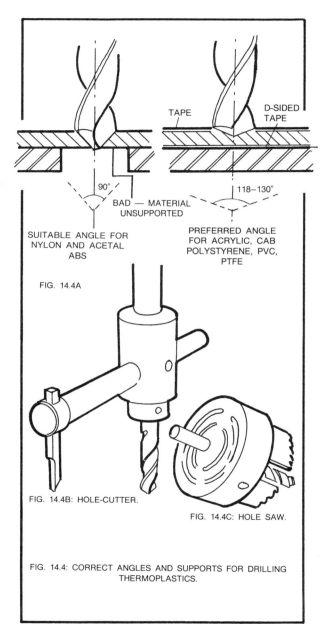

TAPE

D-SIDED TAPE

90°
BAD — MATERIAL UNSUPPORTED

118–130°
PREFERRED ANGLE FOR ACRYLIC, CAB POLYSTYRENE, PVC, PTFE

SUITABLE ANGLE FOR NYLON AND ACETAL ABS

FIG. 14.4A

FIG. 14.4B: HOLE-CUTTER.

FIG. 14.4C: HOLE SAW.

FIG. 14.4: CORRECT ANGLES AND SUPPORTS FOR DRILLING THERMOPLASTICS.

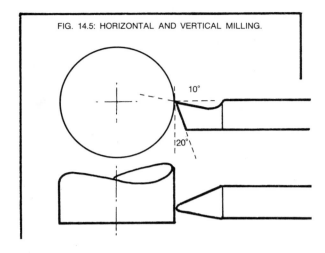

FIG. 14.5: HORIZONTAL AND VERTICAL MILLING.

stressed area. The best way of achieving a low-stress cast block is to make it cure slowly so that the temperature does not rise very much. This can be done by: (1) casting in layers; (2) using a filler powder; or (3) by using less catalyst (seek resin supplier's advice).

Warning: When holes have to be drilled in glass-reinforced polyesters, use tungsten carbide-tipped tools, and plenty of thick soap solution as a lubricant. Wear goggles and a dust mask if any dust is created. When large holes, or large numbers of holes, have to be drilled or cut into mouldings, fine particles of glass fibre will be produced. It is important to avoid inhaling these particles.

TURNING

All standard turning operations can be used on both thermoplastics and thermosetting plastics. The shape of the cutting tool is important, because it is essential that the swarf or waste material leaves the cutting edge immediately it is created. Figures 14.4 and 14.5 shows the recommended tool shape and angles for most thermoplastics; angles for other plastics are given on page 123. Whenever possible, try to achieve a long steady stream of swarf material leaving the cutter. Do not allow swarf to wind round the workpiece and *never attempt to remove it while the chuck is still rotation.* It is usual to have a front clearance angle of 20°, a side clearance of 10° and a top rake angle from 0–10°.

In general, use high speeds and slow feeds, and a lubricant of soap solution or paraffin will usually produce a better finish than if the material is turned dry.

Pressure from the tool often causes the material to 'squash in' slightly, and at the end of

the operation the overall diameter may be oversized. When boring holes the same effect occurs, but causes the hole diameter to be undersized.

When screw-cutting on the lathe, always use the coarsest, deepest thread possible.

If turning operations are to be carried out on plastics materials of small diameters (relative to their length), the work should be supported, because in these circumstances most plastics will flex and, under pressure, move away from the cutting tool. At the end of operations, when it becomes necessary to 'part-off', or separate the work from the stock material in the chuck, it is wise to chamfer back the edges before using a parting-off tool. This prevents burrs occurring on the outside edge of the machined part.

HORIZONTAL AND VERTICAL MILLING: ROUTING

For all these operations it is most important that the plastics workpiece is fully supported, because of its ability to flex under the pressure from the cutter. Double-sided adhesive tape and clamps should be used to prevent the material from lifting and additional packing should be placed on the material surface at all the clamping points to prevent bruising. It is essential that the swarf is removed from the cutter, either by blowing air from a vacuum cleaner at the cutting face or washing waste material away with large quantities of soap solution, paraffin or soluble oil. (Soluble oil stains some materials — for example, white nylon will become discoloured.) End milling is not generally recommended, because these tools have a tendency to clog. Router cutters used for wood machining are preferred, and air blown from a vacuum cleaner should be used to blow the swarf clear, as this operation is usually carried out dry.

Nylon being turned at high speed. Note the continuous length of swarf leaving the cutting tool.

To avoid sharp concave corners, the outside tip(s) should be slightly ground round to create a radius in the machined corner. The base of a flat-bottomed router tool should be ground back at its centre to create a clearance between the cutter and the work surface.

Warning: A router is a very high-speed machine, and this work must be carried out only by your teacher.

Machining operations often cause plastics materials to become stressed around the machined area. When machined edges are to be cemented it is wise to relieve the stress. This can often be achieved by placing the machined part in boiling water and leaving it to cool slowly to room temperature. (But for polystyrene sheet, a lower temperature of 65°C is necessary.) When machined or moulded parts have not been stress-relieved they can develop silvery hairline cracks, especially at cemented joints, which are particularly noticeable in clear materials.

Machine turning PVC bar stock 45mm diameter using soluble oil as a coolant.

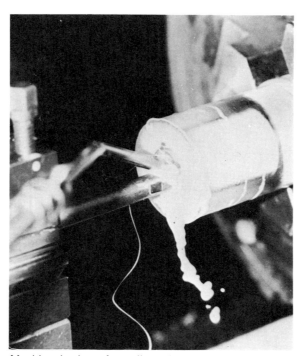

Machine boring of acrylic rod using soluble oil coolant.

Turned laminated PVC trinket box. The laminations are cemented together and machined as a block.

USING STRESS PATTERNS IN ACRYLIC MODELS TO ANALYSE STRUCTURAL DESIGN

Clear acrylic sheet or polycarbonate sheet may be cut out to represent the shape of an engineering product, for example a hook, a spanner, G-clamp, or beam. If pressure is applied to this acrylic model as it might in real life, stress patterns (patterns showing the lines of force) will develop in the model in exactly the same way as they would in the real product. It is possible to see these stress patterns by polarizing a light beam and shining it through the model. Polaroid sunglasses give this effect, and the bright rainbow-coloured stress patterns can easily be seen.

When carrying out this visual stress analysis, one can start with no pressure and increase it to the point where the product is about to break. It is fascinating to see how the patterns change and, as the pressure increases, how other areas become involved. Redesign of the test section soon reveals how even minor alterations can greatly improve the structural capability of the model without adding further material.

This sort of analysis is used for large-span building structures to check whether the model stresses are the same as those calculated. These plastics models are used in many engineering design situations.

STRESSES FROM MOULDING PROCESSES

When a plastics moulding is made, force is used to create the new shape. This applies particularly to sheet-forming processes, to extrusion, injection moulding, and press moulding. The material will move and flow under this force,

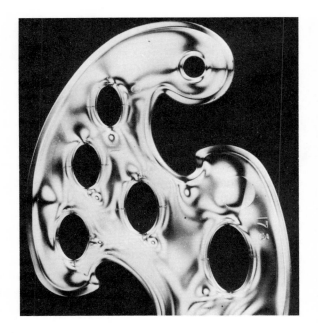

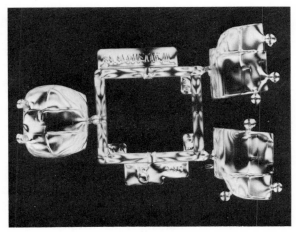

Clear polystyrene mouldings viewed under polarised light. The dark patterning reveals areas of stress concentration (forces frozen into the moulding during manufacture).

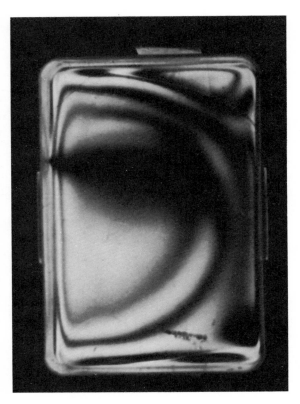

Stress lines show that the box was moulded off centre of the long side.

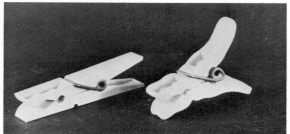

The peg on the right hand side of the picture was dropped into boiling water. The material softened and distorted as the tension in the material released.

causing areas of tension to develop in the moulding when it cools. The material is then said to be in a *state of stress*. This is most likely at points where there are changes of thickness, and where it has been stretched and compressed.

In many cases these stresses will not greatly affect the quality or life of a moulding. But stress *will* affect the performance and life of a moulding if it cannot expand and contract with changing atmospheric temperature or when it has to be used in very cold conditions. If it is used near organic solvents and other chemicals, or if it suffers more forces upon it — that is twisting, bending, dropping — then it may break. The solvents in cements and adhesives can easily attack a moulding under stress and will cause silvery hairline cracks to appear in clear materials (see photo, page 119). Because these stressed areas are a weakness that may affect the life and performance of a moulding, it is often worthwhile carrying out the following process to relax or relieve the stress. The operation is known as normalizing, and is similar to annealing a metal.

In principle it is necessary to take the material up to just below the lowest temperature at which the material would start to soften. Then allow the material to cool down again very, very slowly, so that no part cools any faster than any other. Machined plastics, materials and mouldings treated in this way are less likely to be attacked by chemicals or to be as brittle as mouldings left untreated.

Materials with softening temperatures above 120°C can be placed in a pan of boiling water and left to cool slowly to room temperature. This removes most of the internal forces. Polystyrene mouldings can be put in a pan of water at 65°C and then allowed to cool to room temperature.

This process is suitable for thermoplastics

materials, but unlikely to have much effect on thermosetting compounds, although they also suffer stress problems. The thermosetting compounds have their stress problems reduced by the introduction of fillers and reinforcements in the initial composition of the material, and in some cases by controlled heating and cooling during processing.

Polyester clear casting resin is sometimes referred to as being flexible — and yet it does not bend. This flexibility is achieved by additives which make the resin less brittle.

In the case of polyester casting resin it must be remembered that the chemical cure starts through the action of a catalyst in the middle of the resin mix. This chemical reaction creates heat and is known as an exothermic reaction. The mixture solidifies *before* the temperature rises. As curing takes place, the resin becomes hotter and hotter, expanding most in the middle. This leads to forces being trapped in the casting. When the material is knocked, the casting cracks to relieve these forces.

Where castings are to be machined at a later stage, great care must be taken to ensure very slow cure, to prevent the build-up of heat and trapped stress

Stress cracks showing in extruded acrylic tube. These have developed because fumes were trapped in the tube as plates were cemented on. The fumes evaporated from the cement and attacked the tube walls.

PROJECTS AND ACTIVITIES

(1) It's a very hot day and you are out shopping. As you are passing some parked cars you notice a lady just about to lock and leave her car and that a new compact disc in its polystyrene case is lying on the seat in the sun. What advice should you give her about the CD and why?

(2) You are working on a project that requires two pieces of nylon rod to telescope into one another. You only have sufficient length of one size of nylon rod making it necessary to halve it and bore out one half and to reduce the diameter of the other half. Describe how you would plan your machining operations and any precautions you would take.

(3) The photographs of machine turning on page 116 show the swarf leaving the cutting tool as a continuous ribbon. Why is this desirable and can it cause any difficulties?

(4) If possible obtain a piece of clear acrylic block 15 mm thick × 50 mm × 50 mm. Using the plastics memory technique outlined on page 33 heat it and squash it in a vice and then allow it to cool. Then drill the block with five or six holes of different size. Take great care to clamp the block securely. Drill each hole a little at a time to remove swarf and use plenty of soap solution as a lubricant. Reheat the block and allow it to return to its original thickness. What has happened to the holes, are they are the same? Where will the effects be most noticeable?

(5) Design and make a piece of apparatus for examining small translucent mouldings and machinings in polarised light. Research the requirements thoroughly and work out a method for putting samples of plastics under load. If possible make the light source controllable so that the unit can be used for photographic recording. Provide a diagrammatic instruction chart so that others know how to use the machine easily. Make sure that the electrical wiring conforms to British Standards. Once complete, use the machine. Evaluate it thoroughly and make a full report on your findings. Keep a record of your time and materials expenses and then cost out the machine fully, based on your time at £3 per hour.

(6) Many structures from cathedrals to chairs can be examined for stress distribution when scale modelled in sheet acrylic or polycarbonate, but under load and examined under polarised light. Measure up a cross section through your local church (establish the heights by 'similar triangle' methods) and then cut out the shape carefully in 3 mm thick plastics sheet. Apply a compressive load to the model structure and using two pairs of polarised glasses and a light source to examine all parts for lines of force. Note how these lines move and change as the load changes.

15 *SAFETY*

SAFE WORKING PRACTICE WITH PLASTICS

(1) It is recommended that all secondary school workshops have available for consultation by staff and pupils at least one copy of British Standard 4163:1984, entitled *Health and Safety in Workshops of Schools and Similar Establishments*. Sections 12, 15.2, 15.3 and 17, relating to plastics equipment, adhesives, plastics materials and personal protection, give detailed, straightforward advice on the range of generally available plastics equipment and materials and the precautions that need to be taken when handling them. The same standard also covers all other workshop practices, including machining, disposal of waste and other guidance relevant to the safe use of plastics in schools.

(2) Manufacturers and materials suppliers provide literature on the safe handling of their materials and readily give advice on applications and technical properties.

(3) ESPI (the Education Service of the Plastics Institute) publishes a safety booklet/wallchart, available from ESPI, c/o Department of Design and Technology at Loughborough University, price 50p. The Director of the service is also prepared to answer specific enquiries regarding plastics, and has available a variety of other literature as resource material for schools plastics project work.

(4) While the above three notes and the following details relate to *safe working practice* for *making* components from plastics materials, it must also be remembered that it is the duty of every designer of a product to specify safe materials in relation to the eventual *use* of a product. When choosing a material, consideration must be given to the age of the user of the product, environmental working conditions (the physical and chemical constraints) and structural performance requirements. Before passing the product on to the user it should be tested by a person qualified to pass it as safe. Where possible, foresee misuse or abuse and always aim to make the product failsafe. For example: toys made for babies and young children should be too large to be swallowed, should be made from non-toxic materials, and should not break into small, sharp pieces when hit hard against the floor.

GENERAL HEALTH AND SAFETY NOTES

Almost *all* materials and processes present health and safety risks when incorrectly handled. The range of plastics is no exception, but the nature of the materials and processes involved create some hazards different from those encountered with traditional materials and processing techniques. All plastics must be regarded as chemicals, their individual structures influencing and determining the method of handling, their working properties and the personal risk they present. Do not use unidentified materials.

Dust

Preparation and finishing operations often require plastics to be rubbed-down or abraded. The fine light dust created can float in the air, creating a hazard. It is important to rub down by hand (rather than machine), and to use dampened wet-and-dry abrasive papers. If this is impossible, dust must be collected and extracted every few minutes, using a vacuum cleaner, and pupils are advised to work over a damp cloth and wear a face mask. *Wet or dry dust rags or dirty cleaning cloths must be disposed of by placing in a polythene bag, which should be tied-off and placed in a dustbin out of doors.*

Fumes

Adhesives, solvents, cements, plastisols and resins give off noxious fumes. Adequate ventilation is essential in all areas where these risks are present. A fume cupboard is useful for many small-scale operations involving solvents, cements, plastisols and resin casting, as a means of localized fume extraction.

Large fibreglass mouldings, or many small mouldings adding up to $1m^2$ in area, should be done in the open air or in a large well ventilated room (over $80m^2$) to allow the fumes to disperse. You are allowed to work with glass-fibre and polyester resins only if you are over 12 years of age. When polyester resin is cast, or used in the making of glass-fibre reinforced mouldings, styrene fumes are given off into the atmosphere, which can irritate the throat, lungs, eyes and skin of those nearby. The fumes are also a fire risk. Pupils making GRP mouldings must wear goggles, gloves and an apron, and use the specially prepared barrier creams and hand-cleansing creams available from the materials supplier. Small quantities of MEKP catalyst should be dispensed *only by the teacher* from the calibrated measuring dispensers made for this purpose. Industrial acetone solvent must *not* be used for washing resin from the skin.

Low styrene emission polyester resins are recommended for school use as they give off fumes after the moulding operation is complete.

Shortly after a GRP moulding has been made, the laminating resin will gel and firmly hold the glass fibres in position. Mouldings should be trimmed at this stage (known as the *green* stage) with a sharp knife before the resin has hardened. Trimming that has to be carried out on a fully cured moulding requires a saw, and this will create dust particles of both resin and glass fibre. Face masks (respirators), goggles and gloves should be worn, as these dusts will irritate nasal passages, eyes and bare skin if left unprotected.

Splashes of organic peroxide (MEKP) or catalysed resin in the eyes must be washed out with large quantities of fresh water and medical advice obtained immediately from a doctor since a long-term injury can be sustained.

If during the lay-up procedure the smell makes someone feel dizzy or sick, then they must be assisted to leave the room to obtain fresh air. Ensure that all doors and windows are open to increase ventilation.

Epoxy resins are poisonous, and polyurethane foam liquid agents, when mixed, give off toxic fumes — neither are suitable for use in schools. Some flexible mould-making compounds that require to be cured by chemical reaction give off harmful fumes. Seek specialist advice before purchase in respect of the application and handling methods.

Hot plastics

Many processes involve the use of heat to form mouldings, and others generate heat due to chemical reaction. Heat processes include the use of strip heaters, ovens, vacuum formers, injection, extrusion, dip coating, welding and compression equipment. Materials that become overheated break down chemically, causing a *fume hazard*. Molten thermoplastics can drip, flow or ignite, creating a *fire hazard*. Molten thermoplastics under pressure can spurt from injection, extrusion and compression moulding machines and moulds and nozzles *must be guarded*. Polycarbonate and nylon both absorb

moisture from the atmosphere (i.e. they are hygroscopic) and must be dried thoroughly before being heated, to prevent steam from being generated during hot moulding operations. All sheet hot moulding and curing processes must be carried out wearing heat-resistant leather gloves.

For all hot pressure-moulding operations (injection, extrusion and compression) the operator should wear goggles or a visor and heat-resistant leather gauntlets. Safety guards must be fitted to all such machines.

Remember that molten plastics stick to the skin and because of their high heat capacity can cause serious burns. It is essential that heat-resistant gauntlet gloves are used whenever plastics are moulded with heat.

Hot-wire cutters should *only* be used for cutting expanded polystyrene foam and then only at the lowest temperature that will allow an even cut. The wire must not cause smoke to be generated. This operation must be carried out in a well ventilated area. (Use an extra-low-voltage safety supply, preferably a 6- or 12-volt battery.) *Never* use a hot-wire cutter to cut *polyurethane foam.*

Exothermic (heat generating) chemical reactions

Polyester resin left-overs and large castings are liable to become extremely hot, owing to chemical reactions. Carry out all work in a well ventilated area, preferably outdoors, well away from solvents and other heat-sensitive materials. Left-over catalysed quantities of resin should be spread out over a metal tray in the open air to prevent heat build-up. Before undertaking large castings in polyester resins, seek the advice of the materials supplier to see if it is possible to slow down the rate of resin cure by using reduced quantities of catalyst. Where possible use

filler powders or make the casting in layers (waiting for the previous layer to cool completely before adding the next layer). When cast resin is to be machined, it is essential that curing during casting is prolonged to avoid internal stress concentrations in the casting. If silvery hairline cracks appear, then stop immediately.

Storage

All plastics must be stored in cool, dry, well ventilated conditions. Sheet materials should be stored vertically, tube and rod materials should be supported along their length in racks. Dust and dirt should not be allowed to build up. Liquid plastics, resins, solvents and adhesives should be kept to small quantities. Special separate storage facilities are necessary for catalysts (organic peroxides) and accelerators (cobalt napthanate), used for glass-reinforced polyester work.

PROJECTS AND ACTIVITIES

(1) Some plastics materials are very brittle but can easily be cut using a hand (tenon) saw. Describe any simple precautions you and your neighbours should take during the operation.

(2) You want to drill some holes through 3 mm thick acrylic sheet 150 mm × 150 mm. Is this a potentially dangerous operation and if so how would you make sure that the holes were drilled perfectly without accident?

(3) Some chemicals used for welding and moulding plastics are dangerous. Can you name three and state when they would be used and the precautions that should be taken.

(4) Dust is often created when cutting and working plastics. Is it harmful and can it be made safe easily?

(5) You want to cut two pieces of foam material: (a) expanded polyurethane and (b) expanded polystyrene. Explain what method you would use to cut each material safely and why one method will not suit the other material.

(6) What problems occur when you burn thermoplastics materials? Select four and using small examples (1 cm²) carry out tests in a fume cupboard.

(7) What do you understand by the term 'fail safe'? Can you find a product on the market which in your opinion *does* 'fail safe' and one which does not?

(8) When you go into your school workshop see how many potential dangers you can spot. List them and discuss them with your teacher.

(9) A friend has just been sent to fetch a sealed 25 kg drum of resin from the other end of the workshop. Is this dangerous — how could he or she be hurt? Is there a right and wrong way to lift heavy objects — apart from using a crane?

(10) You are at home and have to repair your glass fibre reinforced canoe. What safety precautions should you take throughout the resin preparation, lay-up operation and cleaning down procedure? Can you name any hazards that might arise?

(11) You are working with a friend using polyester resin to make a casting. Some catalysed resin splashes into his eye. What should you do?

(12) While strip bending some acrylic sheet it catches fire. You must do something quickly. What? Describe your actions.

(13) Make a cartoon drawing for your teacher showing a fellow student working dangerously. How many hazards can you show in the one illustration?

(14) You have been asked to design a pull along toy for a 2 year old child. Do you have to consider any safety factors? Prepare a design and describe the safety considerations.

(15) Very young children love toys that make a noise or have some mechanical movement when towed by a string. Borrow a toy of this type and criticise it from a safety point of view.

(16) Your mother has just told you that you have to look after a young neighbour (2 years old) tomorrow afternoon. Walk round your home and see how many dangerous situations you can find.

(17) A blind person will be visiting your class/school this afternoon. As safety monitor for your form what precautions will you take?

(18) Can you set up a safety committee in your form and challenge the rest of the class to a safety competition?

(19) Why should thermoplastics materials be turned every few seconds when being heated on a strip bender? Would you turn thick (6 mm) sheet more often or less often than 3 mm material? Would you turn a wide sheet in which you wanted to make a long bend more or less often than a narrow sheet of the same thickness?

(20) Do you need to wear gloves when handling hot plastics sheet material?

(21) A friend has pressed a shape in PVC sheet but it has not formed to the depth he requires. He proposes putting the sheet back in the oven and reheating it so that he can continue to press the sheet to the depth required. Is this possible? What will happen?

(22) When you make a sheet blow moulding do you require more or less pressure for a small hemisphere than a large one? It is easy to see which will require the most air.

Table 4 Machining rigid plastics

Thermoplastics materials	Coolants	speed (m/min)	feed rate (mm/rev)	top rake (deg.)	side rake (deg.)	clearance (deg.)	revolutions/minute (small diam.—up to 6 mm)	(large diam.—up to 29 mm)	point angle (deg.)	helix (deg.)	clearance (deg.)
							Turning		**Drilling**		
ABS	air, water, soap solution	150–170	0.3	5	10	15			90	60	10–19
polystyrene	air, water, soap solution	100–300	0.25	2	15	15	1500	100	90–118	20–30	10–15
acrylics	air, water, soap, solution, paraffin	150	0.20	0	15	15	2000	250	118	27	15–20
CAB	air, water, sol. oil for acetate	100–350	0.75	0–5	10	15	6000	500	130	20	10–15
Nylons	air, oil	150–350	0.25	5	0	15	1500	300	90	17	10–15
polyacetal	air, oil	150–350	0.25	3	0	15	1500	300	90	15–25	10–15
polycarbonate	air, water soap solution	200–230	0.3	7	15	7			80	27	10–15
polyethylene	air, water	60–150	0.25	0–5	10	20	1000	250			
polypropylene	air, water	60–150	0.25	0–5	10	20	1000	250			
PVC	air, water, soap solution	100–350	0.75	0–9	10	20	6000	500	130	15–25	10–15
PTFE	air, water	60–150	0.05	0	0	20	1000	250	118	20–30	10–15
Thermosetting plastics											
urea formaldehyde	soluble oil	100–230	0.25	0	15	15	5000	700	118		
phenol formaldehyde	soluble oil	100–230	0.25	0	15	15	5000	700	118		
polyester resin	soluble oil	100–300	0.08	0	15	20	6000	400	118		

NOTE: Coolant air supply should be from the low pressure 'blow' side of a vacuum cleaner. Do not used a compressed air supply.

Table 5 Maximum and minimum environmental working temperatures under load

Material	Maximum temperature (°C)	Minimum temperature (°C)
polyethylene	85	−50
polypropylene	95	−25
polystyrene	70	−50
PMMA	70	−50
PVC	70	−30
polyacetal	90	−50
PTFE	150	−150
Nylon 6	110	−40
Nylon 66	110	−30
polycarbonate	110	−70

Table 6 Oven-moulding temperatures for common thermoplastics sheet materials

Acrylic (PMMA)	160–170°C
PVC	120–140°C
CAB	130–150°C
ABS	130–150°C
polystyrene	130–150°C
polycarbonate	170–190°C
polyethylene	150–160°C
polypropylene	150–160°C
Plastazote	90–100°C
Foamex (PVC)	130–135°C

MATERIALS SELECTOR CHART

Key
LD — low density
HD — high density
HI — high impact

If the product is to be a thin sheet or skin go to **A**.

If a plastic surface finish is required go to **B**.

For **physically** tough working environments go to **C**.

For difficult **chemical** environments go to **D**.

For **exterior** applications go to **E**.

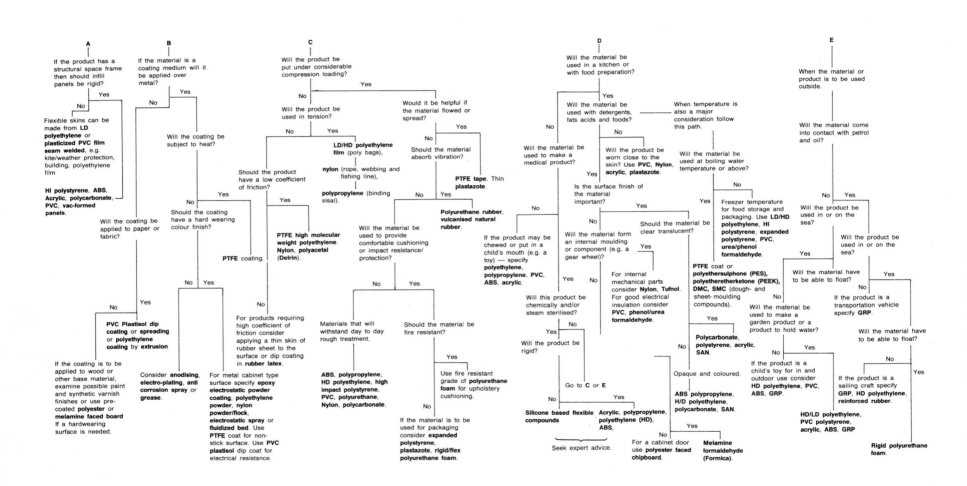

A

If the product has a structural space frame then should infill panels be rigid?

No — Yes

Flexible skins can be made from **LD polyethylene** or **plasticized PVC film seam welded**, e.g. kite/weather protection, building, polyethylene film

HI polystyrene, ABS, Acrylic, polycarbonate, PVC, vac-formed panels.

B

If the material is a coating medium will it be applied over metal?

No — Yes

Will the coating be subject to heat?

Will the coating be applied to paper or fabric?

No — Yes

Should the product have a low coefficient of friction?

Yes — No

Should the coating have a hard wearing colour finish?

Yes — No

PTFE coating.

No — Yes

PVC Plastisol dip coating or **spreading** or **polyethylene coating** by **extrusion**

If the coating is to be applied to wood or other base material, examine possible paint and synthetic varnish finishes or use pre-coated **polyester** or **melamine faced board** If a hardwearing surface is needed.

Consider **anodising, electro-plating, anti corrosion spray** or **grease.**

For metal cabinet type surface specify **epoxy electrostatic powder coating, polyethylene powder, nylon powder/flock, electrostatic spray** or **fluidized bed**. Use **PTFE coat** for non-stick surface. Use **PVC plastisol** dip coat for electrical resistance.

C

Will the product be put under considerable compression loading?

No — Yes

Will the product be used in tension?

No — Yes

LD/HD polyethylene film (poly bags), **nylon** (rope, webbing and fishing line), **polypropylene** (binding sisal).

Would it be helpful if the material flowed or spread?

No — Yes

Should the material absorb vibration?

No — Yes

PTFE tape. Thin **plastazote.**

PTFE high molecular weight polyethylene. Nylon, polyacetal (Delrin).

For products requiring high coefficient of friction consider applying a thin skin of rubber sheet to the surface or dip coating in **rubber latex**.

Will the material be used to provide comfortable cushioning or impact resistance/ protection?

No — Yes

Polyurethane rubber, vulcanised natural rubber.

Materials that will withstand day to day rough treatment.

Should the material be fire resistant?

Yes — No

Use fire resistant grade of **polyurethane foam** for upholstery cushioning.

ABS, polypropylene, HD polyethylene, high impact polystyrene, PVC, polyurethane, Nylon, polycarbonate.

If the material is to be used for packaging consider **expanded polystyrene, plastazote, rigid/flex polyurethane foam.**

D

Will the material be used in a kitchen or with food preparation?

Yes

Will the material be used with detergents, fats acids and foods?

No — No

When temperature is also a major consideration follow this path.

Will the material be used to make a medical product?

Will the product be worn close to the skin? Use **PVC, Nylon, acrylic, plastazote.**

Will the material be used at boiling water temperature or above?

No

Freezer temperature for food storage and packaging. Use **LD/HD polyethylene, HI polystyrene, expanded polystyrene, PVC, urea/phenol formaldehyde.**

Is the surface finish of the material important?

No — Yes

Will the material form an internal moulding or component (e.g. a gear wheel)?

Should the material be clear translucent?

Yes

PTFE coat or polyethersulphone (PES), polyetheretherketone (PEEK), DMC, SMC (dough- and sheet- moulding compounds).

If the product may be chewed or put in a child's mouth (e.g. a toy) — specify **polyethylene, polypropylene, PVC, ABS, acrylic.**

Yes — No

For internal mechanical parts consider **Nylon, Tufnol.** For good electrical insulation consider **PVC, phenol/urea formaldehyde.**

Will this product be chemically and/or steam sterilised?

Yes — No

Will the product be rigid?

No — Yes

Go to **C** or **E**

No — Yes

Silicone based flexible compounds

Acrylic, polypropylene, polyethylene (HD), ABS,

Seek expert advice.

For a cabinet door use **polyester faced chipboard.**

Will the material be used to make a garden product or a product to hold water?

No — Yes

Polycarbonate, polystyrene, acrylic, SAN.

Opaque and coloured.

ABS polypropylene, H/D polyethylene, polycarbonate, SAN.

If the product is a child's toy for in and outdoor use consider **HD polyethylene, PVC, ABS, GRP.**

Yes

Melamine formaldehyde (Formica).

E

When the material or product is to be used outside.

Will the material come into contact with petrol and oil?

No — Yes

Will the product be used in or on the sea?

Yes

Will the product be used in or on the sea?

Will the material have to be able to float?

Yes — No

If the product is a transportation vehicle specify **GRP**.

Will the material have to be able to float?

No — Yes

If the product is a sailing craft specify **GRP, HD polyethylene, reinforced rubber.**

HD/LD polyethylene, PVC polystyrene, acrylic, ABS, GRP.

Rigid polyurethane foam.

PROCESS SELECTOR CHART

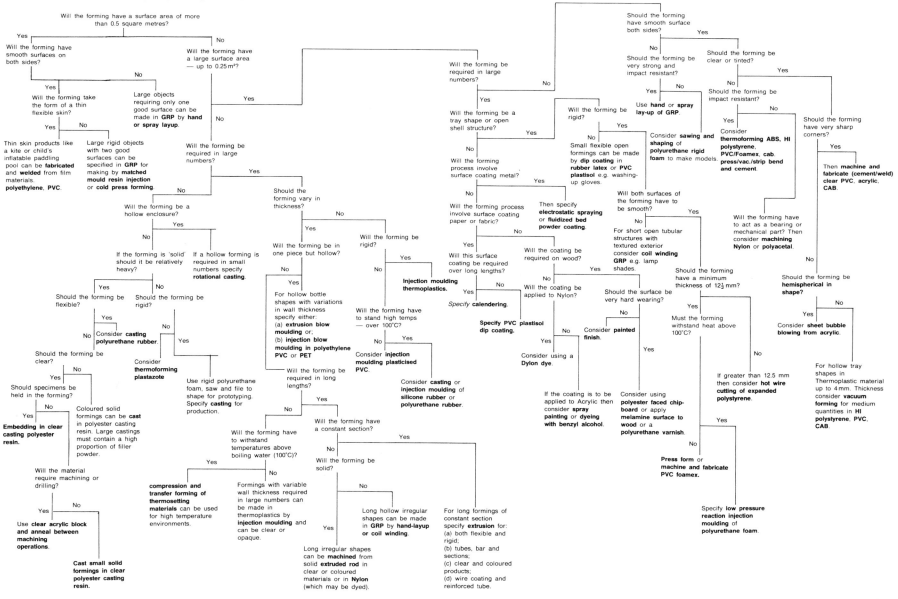

Will the forming have a surface area of more than 0.5 square metres?

Yes — Will the forming have smooth surfaces on both sides?

Yes — Will the forming take the form of a thin flexible skin?

Yes — Thin skin products like a kite or child's inflatable paddling pool can be **fabricated** and **welded** from film materials. **polyethylene, PVC.**

No — Large rigid objects with two good surfaces can be specified in **GRP** for making by **matched mould resin injection** or **cold press forming**.

No — Large objects requiring only one good surface can be made in **GRP** by **hand or spray layup**.

No — Will the forming have a large surface area — up to 0.25 m²?

Yes → **No** → Will the forming be required in large numbers?

No — Will the forming be a hollow enclosure?

Yes — If the forming is 'solid' should it be relatively heavy?
- **Yes** — Should the forming be flexible?
  - **Yes** — Consider **casting polyurethane rubber**.
  - **No** — Should the forming be clear?
    - **No** — Should specimens be held in the forming?
      - **Yes** — **Embedding in clear casting polyester resin.**
      - **No** — Will the material require machining or drilling?
        - **Yes** — Use **clear acrylic block** and **anneal between machining operations**.
        - **No** — **Cast small solid formings in clear polyester casting resin.**
    - Coloured solid formings can be **cast** in polyester casting resin. Large castings must contain a high proportion of filler powder.
- **No** — Should the forming be rigid?
  - **No** — Consider **thermoforming plastazote**.
  - **Yes** — Consider **thermoforming plastazote**.

If a hollow forming is required in small numbers specify **rotational casting**.

Will the forming be required in large numbers?

Yes — Should the forming vary in thickness?

No — Will the forming be rigid?
- **No** — Use rigid polyurethane foam, saw and file to shape for prototyping. Specify **casting** for production.
- **Yes** —

Yes — Will the forming be in one piece but hollow?
- **No** — Will the forming be required in long lengths?
  - **No** — Will the forming have to withstand temperatures above boiling water (100°C)?
    - **Yes** — **compression and transfer forming of thermosetting materials** can be used for high temperature environments.
    - **No** — Formings with variable wall thickness required in large numbers can be made in thermoplastics by **injection moulding** and can be clear or opaque.
- **Yes** — For hollow bottle shapes with variations in wall thickness specify either:
  (a) **extrusion blow moulding** or;
  (b) **injection blow moulding** in polyethylene **PVC** or **PET**.

Will the forming be rigid?
- **Yes** → Will the forming have to stand high temps — over 100°C?
  - **No** — Consider **injection moulding plasticised PVC**.
  - **Yes** — Consider **casting** or **injection moulding** of **silicone rubber** or **polyurethane rubber**.
- **No** — **Injection moulding thermoplastics.**

Will the forming have a constant section?
- **Yes** →
- **No** — Will the forming be solid?
  - **No** — Long hollow irregular shapes can be made in **GRP** by hand-layup or coil winding.
  - **Yes** — Long irregular shapes can be **machined** from solid **extruded rod** in clear or coloured materials or in **Nylon** (which may be dyed).

For long formings of constant section specify **extrusion** for:
(a) both flexible and rigid;
(b) tubes, bar and sections;
(c) clear and coloured products;
(d) wire coating and reinforced tube.

Will the forming be required in large numbers?

No → Will the forming be rigid?

Yes — Will the forming be a tray shape or open shell structure?
- **No** — Will the forming process involve surface coating metal?
  - **No** — Will the forming process involve surface coating paper or fabric?
    - **Yes** — **Specify calendering.**
    - **No** — Will this surface coating be required over long lengths?
      - **Yes** — **Specify PVC plastisol dip coating.**
      - **No** — Consider using a **Dylon dye**.
  - **Yes** — Then specify **electrostatic spraying** or **fluidized bed powder coating**.

Will the coating be required on wood?
- **No** — Will the coating be applied to Nylon?
  - **No** — Consider **painted finish**.
  - **Yes** — If the coating is to be applied to Acrylic then consider **spray painting** or **dyeing with benzyl alcohol**.
- **Yes** — Will the surface be very hard wearing?
  - **No** — Consider using **polyester faced chipboard** or apply **melamine surface to wood** or a **polyurethane varnish**.
  - **Yes** — Consider using **polyester faced chipboard** or apply **melamine surface to wood** or a **polyurethane varnish**.

Yes — Small flexible open formings can be made by **dip coating** in **rubber latex** or **PVC plastisol** e.g. washing-up gloves.

Consider **sawing and shaping** of **polyurethane rigid foam** to make models.

For short open tubular structures with textured exterior consider **coil winding GRP** e.g. lamp shades.

Should the forming have smooth surface both sides?
- **No** — Should the forming be very strong and impact resistant?
  - **Yes** — Use **hand** or **spray lay-up of GRP**.
  - **No** — Will both surfaces of the forming have to be smooth?
    - **No** —
    - **Yes** — Will the forming have to act as a bearing or mechanical part? Then consider **machining Nylon** or **polyacetal**.
- **Yes** — Should the forming be clear or tinted?
  - **No** — Should the forming be impact resistant?
    - **Yes** — Consider **thermoforming ABS, HI polystyrene, PVC/Foamex, cab. press/vac./strip bend and cement**.
    - **No** — Should the forming have very sharp corners?
      - **Yes** — Then **machine and fabricate (cement/weld) clear PVC, acrylic, CAB**.
      - **No** — Should the forming be hemispherical in shape?
        - **Yes** — Consider **sheet bubble blowing from acrylic**.
        - **No** — For hollow tray shapes in Thermoplastic material up to 4mm. Thickness consider **vacuum forming** for medium quantities in **HI polystyrene, PVC, CAB**.
  - **Yes** —

Should the forming have a minimum thickness of 12½ mm?
- **No** — Must the forming withstand heat above 100°C?
  - **Yes** — If greater than 12.5 mm then consider **hot wire cutting of expanded polystyrene**.
  - **No** — **Press form** or **machine and fabricate PVC foamex**.
- **Yes** — Specify **low pressure reaction injection moulding of polyurethane foam**.

USEFUL ADDRESSES

The following list of materials and equipment suppliers' addresses and sources of information is not intended to be comprehensive. Plastic manufacturers and fabricators are to be found in most local home areas and when approached will gladly advise on suitable materials and supply small quantities. Many of the more common plastics are readily available from bigger DIY stores, model shops and local stockists and all are prepared to advise on safe usage.

When seeking help from a company take your work with you so that your contact can see exactly what is needed from your drawings and models. Often you will be given materials and samples to experiment with and an invitation to go back if things don't go according to plan. Don't forget to write and thank them for any help you receive and if possible send a photograph to show your success following their help. Plastics manufacturers are busy people so before you visit a company prepare a list of the questions you want to ask and always obtain the name of the person you are dealing with and telephone for an appointment.

INFORMATION

ESPI Loughborough University of Technology, Ashby Road, Loughborough, LE11 3TU Tel. Loughborough 0509 232065 (*Wide variety of free and purchaseable pamphlets. Further sources of information, booklists, film and video lists*)

BP Educational Service, Britannic House, Moor Lane, London, EC27 9BU (*Detailed catalogue including 'Which Plastics?'*)

SHEET MATERIALS

Amari Plastics PLC, 2 Cumberland Avenue, Park Royal, London, NW10 7RL (*Very wide range plastics sheet materials*)

EMA Model Supplies Ltd, 58–60 The Centre, Feltham, Middlesex (*CAB tube, CAB sheet and other plastics modelling supplies*)

Smith Brothers Asbestos Co. Ltd, Freemans Common Road, off Aylestone Road, Leicester (*Very wide range of plastics materials and safety equipment*)

Robert Horne, Product Promotions, Huntsmans House, Unit 3 Bermondsey Trading Estate, Rotherhythe Road, London, SE16 3LW (*Corrugated plastics sheet*)

Tri-Pack Ltd, Roberts Street, Grimsby, South Humberside, DN32 8AD, Tel 0472 55038/9 (*corrugated plastics sheet*)

Seawhite of Brighton Ltd, 61 Waterloo Street, Hove, Sussex, BN3 1AA (*Polystyrene PVC sheet mats*)

Slaters Plasticard Ltd, Royal Bank Buildings, Temple Road, Matlock Bath, Matlock, Derbyshire, DE4 3PG (*Polystyrene sheet textured for different modelling applications. Polystyrene lettering*)

G. E. Plastics Ltd, Birchwood Park, Risley, Warrington, Cheshire, WA3 6DA Tel. 0925 824232 (*Lexan polycarbonate sheet*)

BXL Plastics Ltd, E.R.P. Division, Mitcham Road, Croydon, Surrey, CR9 3AL Tel. 01 684 3622 (*Plastazote/Evazote*)

Homalocks Ltd, 21 Regal Drive, Soham, Ely, Cambridgeshire, CB7 5BA Tel. 0353 720945 (*Injection moulded fittings for Corriflute polypropylene fluted sheet*)

ADHESIVES

Bostik Ltd, Ulverscroft Road, Leicester

Dzus Fasteners Ltd., Farnham Trading Estate, Farnham, Surrey, GU9 9PL

Evode Ltd, Common Road, Stafford, ST16 3EH

Epoxy (Araldite Resins) CIBA-Geigy Ltd., Duxford, Cambridge

WELDING EQUIPMENT

Parnall & Sons, Lodge Causeway, Fishponds, Bristol, BS16 3JU (*Mini Boy welder*)

Welwyn Tool Co., Stonehills House, Welwyn Garden City, Herts, AL8 6NU (*Hot air welders*)

W. J. Furse & Co. Ltd, Engineering Division, Wilford Road, Nottingham, NG2 1EB, Tel. 0602 863471 (*Electrical heaters and controllers*)

Bielomatik London Ltd, Cotswold Street, London, SE27 0DP (*Welders*)

Goodburn Plastics Ltd, Arundel Road Trading Estate, Uxbridge, Middlesex (*Hot air welders*)

Acru-Electric Tool Manufacturing Co., Acru Works, Demmings Road, Cheadle, Cheshire (*Welding tool*)

VAC FORMING AND SHEET FORMING MACHINES

C. R. Clark & Co. Ltd, Unit 3, Betws Industrial Park, Ammanford, Dyfed, SA18 2LS Tel. 0269 2329

M. L. Shelley & Partners Ltd, St. Peter's Road, Huntingdon, PE18 7HE

Parnall & Sons, Ltd, Plastics Division, Lodge Causeway, Fishponds, Bristol, BS16 3JU

Formech Ltd., 10 Lambton Place, Westbourne Grove, London, W11 2SH Tel. 01 221 4121

SATRA (Shoe and Allied Trades Research Association), Satra House, Rockingham Road, Kettering, Northamptonshire

SUNDRIES

F. Brauer Ltd, Grove Road, Harpenden, Herts., AL5 1PZ (*Toggle Clamps*)

OVENS

R. W. Jennings & Co. Ltd, Scientech House, East Bridgeford, Nr. Nottingham (*Morgan Grundy ovens*)

Griffin & George, c/o Fisons Scientific Apparatus, Bishop Meadow Road, Loughborough, Leicestershire Tel. 0509 231166 (*A variety of plastics equipment and materials*)

INJECTION MOULDING MACHINES

Manual
The Small Power Machine Co. Ltd, Bath Road Industrial Estate, Chippenham, Wilts. Tel. 0249 50366

Dohm Ltd, 167 Victoria Street, London W1

Automatic
Austin Allen, Queens Park Road, Harold Wood, Essex Tel. Ingrebourne 44744

INJECTION MOULDING MATERIALS

Small quantities are often available from local moulders. Polystyrene, polyethylene and polypropylene are most suitable for school ram injections machines.

MOULD MAKING MATERIALS

K & C Mouldings, Spa House, Shelfanger, Diss, Norfolk (*Wide variety mould making materials*)

Boots Chemist (*Plaster: small quantities*)

Davids Isopon, Wellingborough, Northants. (*P38/Car body polyester filler*)

Alec Tiranti Ltd, 70 High Street, Theale, Berks. or 21 Goodge Place, London W1 (*Tools and mould materials*)

Vinatex Ltd, Mill Lane, Carshalton, Surrey (*Vinamold*)

Potters Casting Plaster (*Plaster: large quantities*)

J. W. Phillips Ltd, Pomeroy Street, New Cross, London SE14 (*Pattern makers' equipment*)

POLYESTER RESIN AND GRP SUPPLIES

Strand Glassfibre Ltd, Williams Way, Wollaston, Wellingborough, Northamptonshire, NN9 7PF Tel. 0933 664455

Scott Bader, Wollaston, Northamptonshire (*Large quantities only*)

K & C Mouldings Ltd, Industrial Estate, Ravenstone Road, Coalville, Leicestershire

Trylon Ltd, Thrift Street, Wollaston, Northamptonshire

K & C Mouldings, Spa House, Shelfanger, Diss, Norfolk, IP22 2DF

B.I.P., Popes Lane, Oldbury, Warley, Worcestershire (*Beetle resins*)

PIGMENTS

Llewellyn Rylands Ltd, Balsall Heath, Birmingham 12

Durham Chemical Distributors Ltd, 55–57 Glengall Road, London, SE15 6NQ Tel. 01 639 2020 (*Materials and information for dyeing acrylics and nylons*)

SAFETY EQUIPMENT

Raydek Safety Equipment, Raydek House, Saltley Trading Estate, Birmingham, B8 1BL

Pal Wear Ltd., P.O. Box 144, Sandhurst Street, Oadby, Leicester, LE2 5LW Tel. Leicester 717641 (*Disposable gloves, aprons, overalls and overshoes*)

Rozalex Ltd. Thorncliffe, Chapeltown, Sheffield (*Barrier cream*)

Kerodex Ltd., Innoxa House, 436 Essex Road, London, N1 3PL (*Hand cleaner*)

FURTHER READING ON PLASTICS

Plastics for Schools, P. J. Clarke, Mills & Boon (1976)
Design and Technology of Plastics, R. Millett, Pergamon (1977)
Design with Plastics (*Project Technology Handbook 8*), Schools Council, Heinemann (1974)
Creative Plastics Techniques, Claude Smale, Van Nostrand Co. (1973)
Classic Plastics, Sylvia Katz, Thames & Hudson
The Inventions that Changed the World, Readers Digest Association, Hodder & Stoughton
Plastics Come of Age, Design Council
ESPI Booklets:
Plastics in Contact with Food
GRP Laminates
Plastics in Furniture
Plastics in Agriculture and Horticulture
Plastics Facts and Figures
Plastics in Building
The Production of Polychloroethene
Resources Box for all Ages and Abilities (available from E.S.P.I. includes samples of plastics teaching materials etc.)
Safety in Practical Studies D.E.S. Publication (available from HMSO. Offers positive guidance for safe working practice for the use of plastics in schools.)
Introduction to Extrusion, O and A Level, Paul N. Richardson, Society of Plastics Engineers, U.S.A. (available from Plastics and Rubber Institute, 11 Hobart Place, London, SW1W OHL)
Introduction to Injection Moulding, O and A Level, Clifford I. Weir, Society of Plastics Engineers, U.S.A. (available from Plastics and Rubber Institute, 11 Hobart Place, London, SW1W OHL)
Injection Moulding (available from PPITB, 950 Great West Road, Brentford, Middlesex)
Extrusion (available from PPITB, 950 Great West Road, Brentford, Middlesex)
Children and Plastics, Mary Horne, Macdonald Educational, (1978) (Project work for classroom use available from ESPI)
Plastic Engineering, F.Y.T. booklet No. 16 (Down to earth introduction well illustrated. Selected use for CSE/O Level) Engineering Ind. Training Board, Watford (1977)
School Technology in Action (Examples of work carried out in secondary schools at various levels as a problem solving exercise) EUP/OUP (1974)
Plastics Forming (Production Engineering Series) Macmillan Engineering Evaluations
Modern Design in Plastics, O and A Level, D.P. Greenwood John Murray (1983)
Ideas for Egg Races (and other practical problem solving activities) British Association for the Advancement of Science, 23 Savile Row, London, W1X 1AB
Plastics and the Environment, O/A Level, J. J. P. Staudinger, (Ed.) Hutchinson (1974)
Recycling of Thermoplastic Wastes Distributed by Adam Hilger Ltd, Techno House, Redcliffe Way, Bristol BS1 6NX
The First Century of Plastics M. Kaufman, Morgan Grampian (1970)
Selection and Use of Thermoplastics (Engineering Guide No. 19) P. C. Powell, Design Council (1977)

INDEX